RISK AND SAFETY IN PLAY

Risk and Safety in Play
The law and practice for adventure playgrounds

PLAYLINK

Written for PLAYLINK by Dave Potter

E & FN SPON
An Imprint of Chapman & Hall
London · Weinheim · New York · Tokyo · Melbourne · Madras

PUBLISHED BY E & FN Spon, an imprint of
Chapman & Hall, 2–6 Boundary Row, London SEI 8HN, UK

Chapman & Hall, 2–6 Boundary Row, London SEI 8HN, UK

Chapman & Hall, GmbH, Pappelallee 3, 69469 Weinheim, Germany

Chapman & Hall USA, 115 Fifth Avenue, New York, NY 10003, USA

Chapman & Hall Japan, ITP-Japan, Kyowa Building, 3F, 2-2-1 Hirakawacho, Chiyoda-ku, Tokyo, 102, Japan

Chapman & Hall Australia, 102 Dodds Street, South Melbourne, Victoria 3205, Australia

Chapman & Hall India, R. Seshadri, 32 Second Main Road, CIT East, Madras 600 035

Distributed in the USA and Canada by Van Nostrand Reinhold, 115 Fifth Avenue, New York, NY 2003, USA

First edition 1997

© 1997 PLAYLINK

Typeset by Amy Boyle, Chatham, Kent

Printed in Great Britain by Alden Press, Osney Mead, Oxford

ISBN 0 419 22370 3

∞ Printed on permanent acid-free text paper, manufactured in accordance with ANSI/NISO Z39.48-1992 and ANSI/NISO Z39.48-1984 (Permanence of Paper)

Contents

Contents

Play statement

Every child needs to play freely in order to grow into a healthy, happy, creative and confident adult. It is the responsibility of the community to ensure that each child has access to a range of stimulating play opportunities.

For more than 30 years PLAYLINK has been helping communities to provide for their children's play. We offer advice, information and playwork training. We carry out site audits for adventure playgrounds and design and build play structures to suit playgrounds' individual requirements.

PLAYLINK works to increase awareness of children's play needs and promote quality in services and play facilities.

PLAYLINK
The Co-op Centre
Unit 5 Upper
11 Mowll Street
London SW9 6BG
Tel: (0171) 820 3800

The research, writing and consultation for this publication would not have been possible without the support of the Sports Council. PLAYLINK is most grateful for their grant which has enabled us to produce this code of practice to a standard which playworkers and managers have a right to expect and which we hope will make a significant contribution to the safety and well-being of children at play.

PART ONE

—

Introduction

Opportunities to play are vital for the survival and healthy development of children. The Charter for Children's Play produced by the Children's Play Council states that:

> play is an essential part of every child's life and vital to the process of human development. It provides the mechanism for children to explore the world around them and the medium through which skills are developed and practised. It is essential for physical, emotional and spiritual growth, intellectual and educational development, and acquiring social and behaviourial skills

ADVENTURE PLAYGROUNDS

PLAYLINK believes that the best play provision aims to offer children access to the widest possible range of experience in a setting free from unacceptable risk. Adventure playgrounds are specifically designed to meet children's play needs in this way. They allow the child to explore, manipulate, directly experience and affect their own environment and manage an acceptable level of risk without coming to harm, as well as to experience the pure pleasure of play. Within the adventure playground setting, it is possible to offer opportunities for creativity and imagination where the emphasis is always on the child's choice and control over their own experience.

An adventure playground is defined as an enclosed play area, supervised by playworkers, which is not accessible to children when not supervised. Typically they provide outside space with a variety of environmental modifications and play structures and an indoor space with tables and chairs and materials for crafts, drama, games and so on. Because they are free of charge at the point of entry, are permanently staffed and community based, they provide a familiar environment for children which gives them both a sense of belonging and ownership and the valuable opportunity for making sustained relationships with trusted adults outside the home.

With the exception of specialist facilities for children with disabilities, such as those provided by HAPA, adventure playgrounds do not provide day care.

Playworkers will ensure that there is no suggestion or inference by formal or informal arrangements (for example by agreeing to prevent a child from leaving the site) that these facilities offer a day-care service. However they do take full responsibility for the health and welfare of children while they are on site.

The aim of these facilities is to provide children who use them with access to physical, human and other resources as stimuli for play activities. Play is an instinctive behaviour through which children interact with the world around them. It enables them to explore and experiment, learn skills and develop as individuals. When children experience a wide variety of high-quality play opportunities this development is significantly enhanced. The quality of the child's play experience is linked to the variety of resources available for use in their play, and the way in which those resources are presented and made available.

Adventure playgrounds offer this variety by giving children access to a wide range of physical and human resources. By placing the play needs of children at the centre of the process they are uniquely positioned to facilitate development and growth in children, and therefore they have an extremely important place within the range of play provision that should be available in the community.

They also give children opportunities to learn about making choices and exercising their autonomy. This evolving independence is vital for all children, but particularly for those whose life experiences are limited by factors such as the effects of disability, special needs, poor health, poverty and poor environmental factors.

Adventure playgrounds also provide opportunities for children of all abilities to mix together in an inclusive play environment.

A unique feature of adventure playgrounds is the expectation that children will participate in the development and modification of the facilities and services. This is most striking where children are involved in the construction of play structures for the playground and dens for themselves. Construction provides opportunities for digging, building and creating. The results are often dramatic and add variety and challenge which enhances the play value of the site and promotes a sense of belonging and ownership among the children.

In some cases the playwork skills and understanding to sustain activities that children have traditionally chosen have been lost from adventure playgrounds or abandoned in response to changes in health and safety regulation. Den building is a good example of a rich play activity which has been an inherent part of adventure playground life since their inception but which is now disappearing. This is regrettable since the experience it offers children cannot normally be found elsewhere and when it is found by children on their own the environments are likely to be at best unsuitable and possibly dangerous. Given the objectives of the child's choice and the offer of the widest possible experience through play, the presumption should be that children will be supported in this and similar activities and ways will be found to reduce risk to acceptable levels.

Although these facilities continue to involve children by giving them choice and control of a range of other aspects of playground life, the loss of opportunities to build, dig, and create has taken away from children an experience of interaction with the environment the need for which is as important to them now as when the concept of adventure playgrounds was first proposed.

This handbook will help voluntary groups and local authorities to enrich the environments that they provide for play. It demonstrates how unacceptable risks can be eliminated without reducing the opportunity for challenge and excitement.

A BRIEF HISTORY

The Danish Landscape Architect, Professor Sorensen, was the first to propose the establishment of adventure playgrounds. He spoke of the need for environments, 'sort of junk playgrounds... in which children could create, shape, dream and imagine'. The first facility to take up the ideas of Professor Sorensen was opened in Emdrup, Copenhagen in 1943.

In the same year, a survey in Britain noted the popularity among children of bomb sites as play areas.

After a visit to Emdrup by Lady Allen of Hurtwood in 1946 the concept of 'junk playgrounds' became a reality in London, when in 1948 the first public example was opened by the Voluntary Organisation of Camberwell. Many other voluntarily run projects followed. The term 'adventure playground' was coined to describe these facilities.

In 1954, the National Playing Fields Association provided capital for two new playgrounds, one in Liverpool and one in London, which were very successful until their closure in 1960.

In the 1970s and early 1980s, many adventure playgrounds were established and run successfully in both rural and urban areas around the country with significant concentrations in centres such as Bristol and Newcastle as well as London.

PLAYLINK

In 1962, the London Adventure Playground Association was set up to 'advance the understanding of the educational, social and welfare values of adventure play'. From the four adventure playgrounds which were the founder members of LAPA, by 1977 there were 25 receiving grant aid from the Inner London Education Authority and others, known as 'Play Parks', run by the Greater London Council in London Parks.

In 1969, concerns shared by both LAPA and ILEA about the haphazard way in which adventure playgrounds were being established and run led to LAPA becoming an agent for ILEA. The Authority funded LAPA's full-time fieldwork and training staff. This enabled LAPA to develop and disseminate good practice and standards for local management, playwork and training which were endorsed by ILEA and widely accepted as best practice for adventure playgrounds.

With the ending of ILEA in 1990, LAPA's relationship with the individual playgrounds changed from monitoring and fieldwork to service delivery.

In 1993 LAPA's name was changed to PLAYLINK in recognition of a new wider remit to play service providers beyond London. PLAYLINK continues to promote the original philosophy that gave birth to adventure playgrounds and to develop the concepts and practice standards which lead to high quality in this type of provision for children.

THIS PUBLICATION

This publication is intended as a code of practice for those developing and operating adventure playgrounds, and provides the foundation and detail of a formal safety policy for individual playgrounds.

It is not an Approved Code of Practice as defined within the terms of the Health and Safety at Work etc. Act 1974. However the Health and Safety Executive welcomes the production of guidance by responsible organizations who clearly understand not only the benefits of adventure playgrounds, but also the precautions which could be expected. It accepts that 'industry-led' codes of practice, that is those that have been designed by practitioners themselves are often as good as, and can be better and more effective than, codes imposed from outside.

As a self-regulating code of practice this publication sets standards which would be referred to in the event of a dispute on matters of health and safety.

The recommendations set out in this edition take account of changes in legislation and regulations which have taken place up to November 1996, and the development of good practice which has taken place since it was last revised in 1984.

USING THIS CODE OF PRACTICE

In Part One, Chapters 1 and 2 review the legal framework in which adventure playgrounds operate. After an introduction to the English legal system, there is focus on the two pieces of legislation which have the most impact on their operation; the Health and Safety at Work etc. Act 1974 and the Children Act 1989. A summary of other significant Laws and Regulations is also included.

Chapters 3 to 10 set out how these legal requirements impact on the work of an adventure playground, and provide specific recommendations on good health and safety practices.

In Part Two, Chapter 11 provides summaries of the main piece of Health and Safety Legislation which relates to Adventure Playgrounds, and Chapter 12 provides details of other significant legislation.

An index has been provided which can direct the reader to specific topics, and is useful for quick reference.

In the preparation of this edition, much assistance has been received from individuals and organizations from the adventure playground movement and beyond, and we are greatly indebted to them. A full list of contributors is provided at Appendix 1.

It is important to remember when considering the guidance and advice set out in this publication that the ultimate test for any action is one of reasonableness. A checklist of 'do's and dont's', however conscientiously applied, can never be a substitute for the exercise of skilled judgement supported by reasons. It is necessary to distinguish between what the law says must be done in specific terms, i.e. where there is no discretion, and where it is a matter of judgement as to how to carry out responsibilities.

Judgement should be exercised in the light of the objectives of the adventure playground and the underpinning values as described in the introduction. That is to say, where judgement is exercised it must balance the objective of allowing the children access to experience and learning through play.

FURTHER READING

The following are the main source documents for this introduction:
Brunner, J. (ed.), (1985) *Play, its Role in Development and Evolution*, Penguin.
Ellis, M.J. (1973) *Why People Play*, Prentice Hall.
HAPA Information Sheet 1, *The Benefits of Adventure Play*, undated.
Moyles, J. (1990) *Just Playing*, OUP.
A Charter for Children's Play, (1992) Children's Play Council.
PLAYLINK, *Open Access Play and the Children Act*, undated.
Shier, H. *Adventure Playgrounds*, undated.

Chapter 1

The legal framework

1.1 INTRODUCTION

There is no law which defines how adventure playgrounds should function. There is however a legal and quasi-legal framework within which they must be operated. This framework is based on a combination of Statute Law and Common Law.

This chapter highlights the main components of this legal framework, and identifies how they affect the operation of adventure playgrounds.

1.2 STATUTE LAW

Statute Law is law which has been drawn up and approved by Parliament. Once approved it is published in written form as an Act of Parliament.

Statute Laws can be both Criminal Laws and Civil Laws. There is an important distinction between Civil Law and Criminal Law. Criminal Law is concerned with punishment, and Civil Law is concerned with compensation. Adventure playgrounds come within the framework of both these branches of law.

The main way that adventure playgrounds come within the jurisdiction of the Criminal Law is through the Health and Safety at Work etc. Act 1974 which is enforced through the criminal courts.

There is a list of Acts of Parliament which affect adventure playgrounds below.

An Act of Parliament often contains duties and powers.

Duties or requirements are mandatory and there is no choice about complying. They are usually very general, and are often elaborated in Regulations, which are published some time after the Act of Parliament is passed. Regulations give details of how duties should be carried out. They also are mandatory.

For example, among other things the Children Act, 1989 gives local authorities a general duty to inspect premises where day care for children under eight years of age takes place. The Act does not describe how this should be done. In 1991 Regulations were published by the Department of Health which described how these inspections should be carried out.

Alongside published regulations there is often Guidance. Guidance provides further or additional information on how duties can be carried out. Although guidance is not mandatory, it provides useful information, and could be used to establish whether or not an adventure playground is meeting its legal requirements. In the Children Act example given above, regulations and guidance were contained within the same publication.

Acts of Parliament also describe powers. These give authority for action, but also give discretion as to whether the action is carried out or not. These powers are also often elaborated in regulations.

For example, Children Act Guidance and Regulation Volume 2, empowers local authorities to provide social, cultural or leisure activities in order to promote the upbringing of children by their families, but leaves it to them to decide whether they do it or not.

Although written down, Acts of Parliament often change and evolve. This is particularly true of Civil Law. For example they can be amended by later Acts of Parliament. Also Regulations can be added after the original Act was made Law, providing they are allowed for in the original Act.

For example, in 1993 Personal Protective Equipment at Work Regulations came into force. These Regulations were additions to the Health and Safety at Work etc. Act which was made law 19 years earlier in 1974.

Also, Civil Statute Law can evolve through the interpretation of Judges. For example, if one party disagrees with another over the meaning of a part of an Act of Parliament, a court case may result. If it does, the judgement given at the trial will help clarify and will sometimes develop the meaning of the original Act of Parliament. The precedents set by Judges in this way are called Case Law.

The main Acts of Parliament of concern to providers and organizers of adventure playgrounds are:

1. The Health and Safety at Work etc. Act 1974, and its Regulations:

- Safety Signs Regulations 1980
- Electricity at Work Regulations 1989
- The Reporting of Injuries, Diseases and Dangerous Occurrences Regulations (RIDDOR) 1985
- Control of Substances Hazardous to Health Regulations (COSHH) 1994
- Workplace (Health, Safety and Welfare) Regulations 1992
- Personal Protective Equipment at Work Regulations 1992
- Manual Handling Operations Regulations 1992
- Management of Health and Safety at Work Regulations 1992
- Provisions and Use of Work Equipment Regulations 1992
- Gas Safety (Installation and Use) Regulations 1994

2. Unfair Contract Terms Act 1977
3. The Equal Pay Act 1970
4. The Rehabilitation of Offenders Act 1974
5. The Sex Discrimination Act 1975
6. The Race Relations Act 1976

7. Employment Protection (Consolidation) Act 1978
8. Occupiers' Liability Acts 1957 and 1984
9. Data Protection Act 1984
10. Food and Environmental Protection Act 1989 Part III, the Control of Pesticides Regulations
11. Fire Precautions Act 1971, as amended by the Fire Safety and Safety of Places of Sport Act 1987
12. Consumer Protection Act 1987
13. Children Act 1989
14. Environmental Protection Act 1990
15. Food Safety Act 1990
16. Food Hygiene Regulations 1991
17. The Activity Centres (Young Person's Safety) Act 1995
18. Adventure Activities Licencing Regulations 1996

These provide a framework of requirements and guidelines which should be complied with as a matter of respect for employees and for playground users, and in order to ensure that parents/carers can be confident in permitting their children to use the facilities.

More information on these can be found in the following pages.

1.3 COMMON LAW

Some laws are not laid down in written form by an Act of Parliament, because they have grown up over time and have never been written down. These make up the Common Law.

The most significant of these is the Tort of Negligence. Under this unwritten law each citizen is said to owe a 'duty of care' to each other citizen. If one person breaches that duty (is negligent), and another person suffers loss, damage or injury as a result, then the first person may be required to compensate the second for that loss or injury. The test as to whether negligence has arisen is one of 'reasonableness' (see below).

This duty of care applies to organizations as well as individuals, and therefore to providers of adventure playgrounds.

Because common law is unwritten it is continually evolving. As with statute law, common law can be interpreted by Judges through case law. In fact, Judges' decisions form the basis for interpretation and understanding of common law, as they are written down and therefore act as precedents.

For example, in the case of *Donoghue v. Stephenson* in 1932 a judge described a test to be used when assessing whether a breech of a duty of care had taken place. He said:

> you must take reasonable care to avoid acts or omissions which you can reasonably foresee would be likely to injure your neighbour.

This test, based on the foresight of the reasonable man (*sic*), is applied in all cases where there is a claim in negligence. For cases where children are involved this has been further defined, and the test in these cases is based on the foresight of a reasonable parent. This is known as the reasonably careful parent test.

1.4 BRITISH STANDARDS

British Standards are another important aspect of the legal framework within which adventure playgrounds operate, and there are British and other Standards which relate to fixed-equipment playgrounds.

As these Standards relate to the provision of unsupervised play, they do not apply directly to adventure playgrounds. However they provide a useful reservoir of information and advice in respect of the design and operation of these facilities, which would almost certainly be referenced alongside this publication in the event of a dispute at common law or statute law. Although compliance with these Standards would not itself confer immunity from legal obligations, they can be used to establish whether reasonable care has been taken and form part of the legal framework in which adventure playgrounds operate.

British Standard (BS) 5696 is concerned with play equipment that is intended for permanent installation outdoors, and gives advice and guidance on the design, construction, performance, installation and maintenance of play equipment and surfaces; and the design and layout of play areas. The German DIN Standard 7926 provides information on items such as aerial runways. They will be superseded by a new European Standard which is currently in draft form.

1.5 CODES OF PRACTICE

Codes of Practice are an additional resource for information on good practice on adventure playgrounds.

Although there is no Approved Code of Practice as defined within the terms of the Health and Safety at Work etc. Act 1974, this publication and other 'industry led' codes of practice (for example the Code of Practice for Safety in Indoor Adventure Play Areas) would certainly be referenced in the event of a dispute or claim under common or statute law. The Health and Safety Executive accepts that such codes of practice which have been designed by practitioners themselves are often as good as, and can be better and more effective than, codes imposed from outside.

1.6 FURTHER READING

The following is the main source document for this chapter:
Keenan, D (1992) *English Law*, Pitman.

Chapter 2

The Health and Safety at Work etc. Act 1974 and the Children Act 1989

2.1 INTRODUCTION

The Health and Safety at Work etc. Act 1974 and the Children Act, 1989, are the major pieces of legislation which inform good practice in the health and safety operation of adventure playgrounds. Conformity to their requirements is mandatory and enforceable.

2.2 THE HEALTH AND SAFETY AT WORK ETC. ACT 1974

2.2.1 Introduction to the Act

One of the principal objectives of the Health and Safety at Work etc. Act is to involve everybody at the workplace – management and workers – in consideration of good health and safety practice and to generate awareness of its importance.

Upon publication it became the co-ordinating legislation for previous statutory provisions relating to the health, safety and welfare of employees while at work. In addition it extended statutory protection to all members of the general public in so far as they might be affected by work activities.

Under the Act duties to uphold reasonable practice in relation to health, safety and welfare fall upon:

- employers with regard to employees;
- employers with regard to the general public;
- employees with regard to fellow employees and themselves;

- employees with regard to the general public;
- the controllers of premises;
- manufacturers and suppliers;
- the self-employed.

It requires all persons at work and all persons responsible for any part of any work environment not only to be aware of the need for health and safety but also to ensure that, as far as is reasonably practicable, optimum conditions exist for minimizing risks arising from health and safety hazards.

Failure to discharge one's duty under the Act is a criminal offence.

2.2.2 Who is responsible?

The primary responsibility for occupational health and for avoiding accidents rests with those who create the risks. This is why the Act directs it message first and foremost to the employers and employees in the workplace.

It is for the employer to ensure that appropriate health and safety policies are in place and that the procedures which flow from these policies are being carried out.

In practice there is a variety of arrangements for the management of adventure playgrounds and the employment of staff. These include playgrounds operated by local authorities or by voluntary organizations and those where arrangements are shared.

Where the operation of the playground is the sole responsibility of one agency, it is the duty of that agency to ensure that the provisions of the Act are implemented. Where responsibility for the operation of the playground is divided between agencies, a formal agreement specifying the particular areas of responsibility of each party should be established.

The health and safety of adventure playgrounds operated by local authorities is assessed by the Health and Safety Executive. Those sites where a local authority has no control of the operation are assessed by the Environmental Health Department of the local authority. A list of Health and Safety Offices is provided at Appendix 2.

Inspectors have wide-ranging powers. They may forbid the carrying out of specific activities within all or part of any place of work (including closing a site), and require improvement in conditions or in practice.

2.2.3 Duties and responsibilities

The Act sets out duties and responsibilities for employers and employees.

Employers must, as far as is reasonably practicable, provide a safe working environment and inform, instruct, train and supervise employees on all matters outlined in the Act.

The Act lists five particular duties of an employer to an employee:

1. the provision and maintenance of plant (equipment) and systems of work that are safe and without risks to health;
2. arrangements for ensuring safety and absence of risks to health when using, handling, storing and transporting articles and substances (this covers everything used at work and all work activities);
3. the provision of relevant information, instruction, training and supervision;
4. maintaining any place of work under the employer's control in a safe condition and without risks to health;
5. the provision and maintenance of a safe working environment without risks to health, and of adequate welfare facilities.

The term 'employee' as used in the Act refers to an individual who works under a contract of employment. The duties of employees are:

1. to co-operate with their employer so far as is necessary to enable the employer's legal duty on health and safety to be carried out;
2. not intentionally or recklessly to interfere with anything provided which would effect the implementation of their employer's legal duty on health and safety;
3. not intentionally or recklessly to interfere with anything provided for their health, safety and welfare;
4   not to interfere with or misuse equipment, plant or machinery in any dangerous way.

2.3 THE MANAGEMENT OF HEALTH AND SAFETY AT WORK REGULATIONS 1992

Some of the requirements of the Health and Safety at Work etc. Act have been elaborated through the issuing of a series of Regulations, which are summarized in Part Two of this publication.

The most significant of these is the Management of Health and Safety at Work Regulations, 1992. These establish the principle of risk assessment which is now central to Health and Safety Management. The Regulations are supported by an Approved Code of Practice which has been drawn up to give practical guidance on carrying out its general duties.

They require employers to make what is called a suitable and sufficient assessment of all the risks arising from their undertaking. This does not need to be complicated or perfect. Checking hazards is mostly a matter of common sense based on experience of working in the playground environment. The assessment must consider risks to both employees and to other persons including members of the public. The purpose of the assessment is to identify the measures that the employer needs to take in order to comply with their legal obligations under the Health and Safety at Work etc. Act and its Regulations.

Where an employer has five or more employees, the significant findings of this assessment and any group of employees identified as especially at risk must be recorded. PLAYLINK recommends that assessments should be recorded irrespective of the number of employees.

2.3.1 Risk assessment

A suitable and sufficient assessment of risks involves identifying hazards and then evaluating the extent of the risk involved. It must identify significant risks, identify and prioritize the measures that need to be taken in order to comply with any relevant legal requirement. It should be appropriate to the nature of the activity on the playground and keep in mind the objective of meeting childrens' play needs, including the need to encounter some acceptable level of risk.

Assessment must be reviewed, and if necessary modified, to take account of any change in circumstances.

There are no fixed rules about how a risk assessment should be undertaken. However the general aims are to:

- ensure that all relevant risks are addressed;
- address what actually happens in the workplace or during the work activity;
- ensure that all groups of employees and others who might be affected are considered;
- identify groups of workers who might be particularly at risk;
- take account of existing preventative or precautionary measures.

The following procedures would provide a suitable framework for risk assessment on an adventure playground:

1. Identify and list all aspects of the **building, outside area, structures and programme** which may give rise to hazards.
2. Identify and list any and **all hazards** to health and/or safety which may result from each of these areas of work.
3. Identify and list any and **all the risks** each of these hazards pose to employees and other persons.
4. Identify and list any and all of the **precautions already taken** in respect of each of the risks identified.
5. Assess the **level of risk which remains** after the precautions identified have been taken (more information is given below on how to do this).
6. Identify and **carry out measures** which are required in order to eliminate, minimize or reduce the risk.

The assessment of the level of risk in step 5 above must take into account both how serious an injury could be and how likely it is to occur. So for example:

- a high risk – would be a very serious injury which has a high probability of occurring;
- a medium risk – would be a very serious injury which is unlikely to occur, or a lesser injury likely to occur;
- a low risk – would be where any injury is unlikely and would be slight if it did occur.

It is important to base a judgement about likelihood of occurrence on experience. For example, the fact that a playground has skilled supervision is a factor in reducing the likelihood of accidents. Use accident records to assist in these judgements.

One way of establishing how serious an injury could be is to use an arithmetical scale, as follows:

5 points   very high, potential for multiple death and/or widespread destruction

4 points   high, causing death or serious injury to an individual, i.e. reportable accidents under the RIDDOR Regulations (see Chapter 9)

3 points   moderate, causing injury or disease capable of keeping an individual off work (school) for three days or more, which may require reporting under RIDDOR Regulations

2 points   slight, causing minor injuries, i.e. person able to continue after first-aid

1 point   no risk of injury or disease

Using the same idea to rank the likelihood of the event occurring, the following could be used:

5 points   very likely, almost certain

4 points   likely to occur, i.e. easily precipitated, with slight carelessness, from external event

3 points   quite possible, the accident is only likely to occur with help, i.e. if somebody slips, failure to replace a light etc.

2 points   possibility/probability is low or minimal

1 point   not likely at all, there is really no risk and accidents will only occur under freak conditions.

By multiplying the rankings for how serious an injury could be by the rankings for the likelihood of the event occurring, it is possible to establish a numerical value representing the level of each risk assessed. Using the scale outlined below these could be categorized as follows:

values 1–8   would represent a low risk

values 9–17   would represent a medium risk

values 18–25 would represent a high risk.

Risks identified as high or medium require the identification and undertaking of measures to eliminate, reduce or minimize the risk.

A record must be kept of each risk assessment which must include:

- the names of those involved in the risk assessment, and the date of the assessment;
- details of the work activities, undertakings and facilities assessed;
- all hazards identified;
- all the risks inherent in each of these hazards and who is particularly at risk;
- precautions already taken;
- the level of risk which remains;
- measures required to eliminate, reduce or minimize the risk.

PLAYLINK recommends that the record should also show what action has been taken as a result of the risk assessment, and when.

A model risk assessment process is provided at Appendix 3.

2.3.2 Preventative and protective measures

Once a risk has been identified the preventative and protective measures that have to be taken depend upon the relevant legislation. In deciding on the measures to be undertaken the following principles must be applied:

1. Avoid the risk altogether if possible.
2. Aim to combat the risk at source.
3. If possible adapt work to the individual to diminish risk.
4. Take advantage of technological progress.
5. Give priority to those measures which protect the whole workplace.
6. Ensure workers understand what they need to do.
7. Promote an active health and safety culture.

The Regulations require arrangements to be in place for the effective planning, organization, control, monitoring and review of the preventative and protective measures.

2.3.3 Information for employees

The risk assessment will highlight information which has to be provided to employees in order to ensure their health and safety.

This information must be capable of being understood. Information should be provided in whatever form is most suitable, taking account of the levels of training, knowledge and experience of employees.

2.3.4 Training

These Regulations require employers to ensure that employees are provided with adequate health and safety training, on recruitment and whenever there are changes which result in an increased or different exposure to risks. It should be repeated where appropriate, be adapted to reflect changes and take place in working hours.

2.3.5 Temporary workers

Where temporary workers are employed there is an additional requirement for employers to provide information on the level and type of qualification and/or skills required in order to do the work required.

2.3.6 Employees' duties

These Regulations require employees to correctly use all work items provided by their employer, in accordance with their training and the instructions they receive, to ensure safe use.

Employees are also required to report any shortcomings in the health and safety arrangements, even when no immediate danger exists, so that employers can take such remedial action as may be required.

2.4. SAFETY POLICIES, REPRESENTATIVES AND COMMITTEES

The Act requires every employer, except those with five or fewer employees, to prepare, revise and bring to the notice of their employees, a written statement covering:

- the general policy of the employees with respect to health and safety at work;
- the organisation and arrangements for carrying out that policy.

While employers with five or fewer employees may communicate their policies verbally, direct to their employees, PLAYLINK recommends that every adventure playground has a written statement of its Health and Safety policy and practices irrespective of the number of employees.

The Health and Safety Executive attaches the greatest importance to this aspect of the Act. For every employer, the safety policy is the basis on which the entire health and safety of the organization and activity will depend. This safety policy must, therefore, be drafted clearly and concisely, enabling the management and workers of the playground to understand their particular duties and responsibilities.

It is recommended that every adventure playground nominates a member of staff, who should normally be the senior worker, to keep the written health and safety policy and practice document under review and advise management about necessary changes. They must call on outside expert advisers such as PLAYLINK to help them where necessary.

2.5 THE CHILDREN ACT 1989

2.5.1 Introduction

The Children Act 1989 came into effect on 14 October 1991. It established a new duty on local authorities to promote the upbringing of children in need by their families, so far as is consistent with their welfare. One particular service is given prominence in the Act, day care and supervised activities for pre-school and school-aged children outside school hours and in the holidays.

The Children Act requires those who provide these schemes to register where children under eight participate.

Registration is of a 'fit person' operating in 'fit premises' and covers:

- qualifications, training and suitability of staff and ratios of staff to children;
- size, suitability, safety, and hygiene of premises;
- requirements for quality of care, numbers of children attending and types of activities;
- suitability and safety of furniture and equipment.

Open access schemes such as adventure playgrounds are required to be registered if they cater for children under eight, and are open for more than two hours at a time on six or more days a year.

Guidance and Regulation Volume II provides details of regulations and guidance concerning the implementation of these aspects of the Act.

2.5.2 Children in need

The Children Act identifies the concept of children in need and defines a child in need as:

> he is 'being unlikely to achieve or maintain, or to have the opportunity of achieving or maintaining, a reasonable standard of health or development without the provision of services by a local authority under this Part' (Part X of the Act)

> his health or development is likely to be significantly impaired, or further impaired, without the provision for him of such services;

> or

> he is disabled.

It requires local authorities to define further and make an assessment of needs, and to identify and plan for services required. Such services can include day care provision and local authorities should be facilitators as well as providers of services.

2.5.3 Day-care services

Introduction

The Act requires local authorities to devise a policy for providing day care for children in need. In establishing that policy they are required to fully consult. The views of parents and children must be fully considered. The policy must allow for a variety of day care facilities. For example:

> day nurseries, playgroups and childminding, out-of-school clubs and holiday schemes, supervised activities, parent and toddler groups, toy libraries, drop in centres and playbuses.

Local authorities have power to provide, regulate and review day care services for children in need, and are required to oversee and co-ordinate functions.

The registration and inspection of day-care services is the main aspect of the regulatory function. The main purposes of registration are to protect children, to reassure parents, to ensure standards compliance and to ensure services are provided within a framework. They can also provide day-care facilities for children who are not in need. The Guidance and Regulations on day care applies principally to children under eight years of age.

Application to adventure playgrounds

Day care services for school age children are described in Children Act Guidance and Regulation Volume ll as follows:

> Day-care Services provide play opportunities for school age children outside school hours and in the holidays in three types of setting, which may overlap; a care setting where children are looked after by other adults when the parents are not available; an open access or drop-in play setting where children go to meet other children and where there is some adult supervision; a special interest setting where children develop particular skills and knowledge.

Adventure playgrounds fall within the category of open access or drop-in play settings.

Open access facilities are differentiated from other care services in the guidance and regulations as follows:

> The important distinguishing feature of open access facilities is that there is no limit on the numbers of children who may attend, and the providers do not undertake to keep the child until he is collected ... open access facilities need to be organised so as to prevent this younger age group coming to harm and to ensure that they are well catered for.

This distinction is important in the context of the registration of day care provision, which is discussed below.

Open access does not imply a lower level of care by playworkers. In both open and closed access schemes the providers have legal responsibility for the safety and care of children present and taking part in activities, as far as is reasonably practical. The difference lies at the point when this responsibility ceases. For adventure playgrounds the full responsibility for the care of children ceases when they leave the site. There must be no implied or explicit understanding or agreement that responsibility extends beyond that point. Explicit information must be provided for parents/carers, by way of a leaflet and/or otherwise concerning this limit to the adventure playground's responsibilities for children who use the facility.

This does not mean to say that playworkers have no responsibility in respect of children who leave. In some circumstances, in the interest of the welfare of the child, steps may need to be taken to establish whether there is any greater than normal risk to which the child may be exposed as a result of choosing to leave the site. The playworker will need to consider factors such as the age and ability of the child, external factors which may affect his/her safety, variations from normal patterns of behaviour and local information and knowledge, for example if a particularly vulnerable child were to decide to leave the site alone.

Action could include checking with the child and/or parents/carers or providing information to parents/carers at a future time. These actions must not be done in a way which implies that the adventure playground accepts full responsibility for children who leave. No undertaking must be given to parents/carers which implies that children will be prevented from leaving.

2.5.4 Standards and quality

The Guidance and Regulations identify six principle standards which should underpin day care provision and which would apply to adventure playgrounds.

1. Children's welfare and development are paramount.
2. Children should be treated and respected as individuals.
3. Parents'/carers' responsibility for their children must be recognized and respected.
4. Cultural, racial, religious and linguistic values should be respected.
5. Parents are generally the first educators.
6. Parents should be able to make informed choices.

In meeting those standards, providers of day-care facilities are required to offer activities which are appropriate to the age and development stage of the children, are flexible and offer variety, involve children in planning, and give opportunities for children to rest.

The Guidance and Regulations emphasize the importance of training for staff, equal opportunities policies and practice and parental involvement. They also underline the importance of attention to health issues, including first-aid training for at least one member of the staff on duty, with all staff having working knowledge of first-aid, food handling and monitoring of the health of staff.

Prominence is given to the need for quality of care, although this is not defined in the Act. Quality issues to be considered include developmental needs and rights of children, the provision of opportunities for learning and socializing with adults and children, freedom from discrimination, rights or expectations of parents and rights or expectations of workers. In its publication *Open Access Play and The Children Act*, PLAYLINK describes quality provision in terms of the application of values and principles.

It is good practice to publish the values and principles of an adventure playground in the form of a leaflet/handbook. This can be in two parts. The first to set out the aims and objectives of the adventure playground, details of how these are achieved through the operation of the facilities, and the policies on behaviour etc. which are in place. The second part to advise parents/carers of what is expected of them and why, and to seek basic information about the children who are users, which will assist the scheme in catering for them.

2.6 REGULATIONS FOR THE REGISTRATION OF DAY CARE FACILITIES

2.6.1 Introduction

Local authorities have a regulatory function in respect of day care facilities. This is discharged principally through the registration of persons and premises. Registration is based on an assessment of how premises used for day care, and the persons providing it, meet the regulations set out in Guidance and Regulations Volume II.

It has been acknowledged by the Department of Health that it is important that the registration system for open access schemes, such as adventure playgrounds, is based on the purposes set out in the Guidance, but does not seek to go further in ways that will pervert the objectives.

The requirements of the Children Act in respect of day-care facilities are principally concerned with the welfare of children under eight years of age, and there are many requirements concerning standards for this age group. However little information is provided concerning the welfare of older children. Adventure playgrounds which embrace and welcome the principles of the Children Act will be adopting an approach to the operation of their service which can apply to children throughout the age range they work with, thereby providing an environment which is appropriate to the welfare of older, as well as younger, children.

The Regulations and Guidance concerning registration in Volume II provide for the use of discretion by local authorities in a range of areas, through use of terms such as 'appropriate', 'suitable', 'sufficient' and so on. For example the Regulations say that premises must be 'suitably clean' and that heating systems must be 'appropriate', without defining what appropriate or suitable mean in these contexts.

These terms have been used throughout this publication to refer to advice and guidance from the Regulations. In order to retain the flexibility they have frequently been left undefined. However this publication as a whole provides a framework of information, advice and good practice within which those involved in the Registration process will be able to establish definitions which are acceptable in the context of the facilities being registered.

2.6.2 Specific requirements which apply to adventure playgrounds

Numbers of children, and ratios of staff to children

The Act specifies (Part X, Section 72 {2}{a}) that there must be an agreed maximum number of children, which represents the capacity of the adventure playground. Considerations to be taken into account in determining the policy on maximum number include:

- the layout of the indoor and outdoor areas;
- how effectively the adventure playground can be supervised by the staff team with larger numbers of users;
- the likely age distribution of the users, and the size of the groups in which they will be playing;
- the kind of activities involved;
- the kind of activities that are likely to be available.

Once established, deciding when the capacity has been reached requires the consideration of a number of factors. These include the type of activity taking place at the time, the resources available and the age range of children on site.

Special procedures should be developed to ensure the safety and satisfaction of 5–7 year olds present when the adventure playground becomes crowded. A staff/child ratio of 1:8 for children under 8 years of age should be used as a guideline, and is not a requirement. Levels of staffing must ensure that the 5–7 year-olds are not overwhelmed by older children. This might mean ensuring adequate numbers of staff to organize activities targeted at the younger children. See PLAYLINK publication *Open Access Play and the Children Act.*

Playworkers must not feel obliged to turn children away in order to conform to a strict interpretation of this ratio, particularly when doing so may endanger children. A higher ratio may be necessary when children with disabilities attend.

Staffing

The Guidance and Regulations requires providers of day care facilities to ensure that at least half the staff hold a relevant qualification, and that the person in charge be qualified unless s/he has considerable experience.

There are now recognized National Vocational Qualifications in playwork at levels 2, 3, and 4. At June 1995, assessment centres for those wishing to qualify had been approved in 34 locations around the UK. However, it will take some time for this system to make any significant impact on the number of playworkers who are specifically qualified to work in such settings as adventure playgrounds. PLAYLINK recommends that the underpinning values, and the competencies, described in the NVQs be used as a guide in assessing whether a worker has the appropriate skills for the job. Other recognised qualifications, e.g. in youth work or early years education, might be relevant in particular cases but their relevance would need to be demonstrated.

Senior workers must have an absolute minimum of two years experience of work with children, and an ability to demonstrate that they have the competence to carry out the range of complex tasks involved in the adventure playground environment.

Premises and space standards

The Guidance and Regulations set out a standard for space based on a formula of 2.3m² of clear space per child.

The application of this space standard to adventure playgrounds is inappropriate. A more reliable way of providing a safe environment for children using an adventure playground, particularly the safety of the youngest users, is the playwork policy and practice. This will take account of the layout of the facilities and the staffs' capacity to supervise them effectively, together with knowledge of how the facilities are used by different age groups. An agreed health and safety policy, based on the undertaking of risk assessments, will provide a safety framework more reliable than the application of a space formula.

There are, in addition, requirements concerning the facilities available at premises where day care takes place. These include: office space, space for providing snacks or main meals, toilets, separation of quiet and boisterous activities (where possible) and outside playspace.

An assessment of the suitability of premises will also consider:

- whether access from the site to the road presents a hazard;
- outside playspace safety;
- glass doors (safety glass or protective plastic film should be used);
- the safety of fires, electric sockets, windows and floor coverings;
- washing, toilet facilities and hygiene;
- cooking facilities and safety in the kitchen area;
- arrangements for keeping the premises clean;
- fire fighting equipment, fire precautions and exits and evacuation procedures.

Furniture and equipment

Guidance and Regulation require that furniture and equipment must be in reasonable condition and repair, sufficient in number, and create a friendly and informal atmosphere. Reasonable age and stage appropriateness and British Standards compliance (where this exists) will be assessed. Materials should be provided for art, collage, sport and games, dressing up, music, jigsaws, construction toys and crafts. In non-domestic premises, kitchen equipment must comply with environmental health regulations.

The principles and standards set out in this publication provide more detailed guidance on the need for, provision, use of and maintenance of these items, which include playground structures.

Reports and records

The Guidance and Regulations require basic records to be kept of the children who use the facilities for day care and the people employed on the management group. Childrens' records are required to contain information on:

- age
- name
- name of parents/carers
- emergency telephone numbers
- information on health problems and (any) medication.

Further information which it may be appropriate to keep includes:

- second emergency contact number;
- dietary requirements;
- school/headteacher/teacher/social worker/other agency;
- position in family.

Any records kept on computer are subject to the Requirements of the Data Protection Act 1984 (see Chapter 12).

Fit persons

Registration of day care facilities under the Children Act also entails an assessment of the persons who provide it and other people with substantial access to children in a day-care setting.

A fit person in this context is someone who has appropriate experience, qualification and ability and the physical and mental capacity to do the job. These are described fully in Regulation and Guidance Volume II, and are referred to in the text of this Publication.

Visits and outings

The Guidance and Regulation makes specific reference to visits and outings from day care facilities. It requires that:

- visits must be carefully planned with additional people if required;
- the use of consent forms is commended.

A more rigorous approach to this aspect of the health and safety of adventure playgrounds is set out in Chapter 10.

2.7 THE REGISTRATION PROCESS

The process of registering an adventure playground under the Children Act will vary according to the locally agreed arrangements.

All will involve an initial assessment of the premises, which may be followed by visits from specialists, for example from the fire service or the Environmental Health Department.

Those who have primary responsibility for health and safety must take part in the inspection process. They must ensure that they are prepared for these visits, and must accompany the registration officers as they undertake the assessment. This will ensure that registration officers are aware of the context in which day care is provided on an adventure playground. If there is any confusion concerning the registration of this type of open access facility, those involved should read the PLAYLINK publication, *Open Access Play and the Children Act,* which is recommended by the Department of Health as the appropriate guidance for registration of adventure playgrounds.

2.8 FURTHER READING

The following are the main source documents for this chapter:

Dennis, M. (1992) *Tolleys Health and Safety at Work Handbook*, ROSPA.

Department of Health, (1989) *The Care of Children, Principles and Practice in Regulations and Guidance*, HMSO.

Health and Safety Commission, (1990) *A Guide to the Health and Safety at Work Act 1974*, HMSO.

PLAYLINK, *Open Access Play and the Children Act*, undated.

South Glamorgan County Council, *Review of Day Care and Related Services*, undated.

The Children Act Guidance and Regulations, Vol. 2, Family Support, Day Care and Education Provision for Young Children, HMSO, 1991.

The Children Act Guidance and Regulations, Vol. 8, Private Fostering and Miscellaneous, HMSO, 1991.

Chapter 3

Staffing and training

3.1 STAFFING

Sufficient well-qualified/trained full-time permanent staffing is essential to the concept of adventure playgrounds, as the success of any adventure playground will depend upon the quality and effectiveness of workers and the trust built between workers and children.

Adventure playground work is very demanding. A wide and detailed knowledge of a range of practical skills is required, alongside awareness and understanding of a range of statutory requirements and their applications to playwork.

However, that knowledge is itself only a background to playwork. The function of playwork in adventure playgrounds is to facilitate the growth and development of children. The playworker is the orchestrator of the resources needed to enhance that development. Therefore understanding and the skilful application of the knowledge of child development, play and children's behaviour is required. Alongside those the playworker will need to be skilful in a number of practical activity areas.

Playworkers are primarily workers with children and it is this function which will place greatest stress upon them. The potential for high levels of demand upon the worker must not be overlooked. Systems for the support of workers including non-managerial supervision must be provided. Managers, in consultation with workers, should set objectives and attainable targets.

The adventure playground is a unique style of workplace. The playworker, though primarily a worker with children and young people in the broad range of physical and social activities we characterize as 'play', is also charged with responsibility for the development and maintenance of the physical environment. The playworker, therefore, is expected to have high levels of skill in two separate and different general areas, each of which may, in their turn, comprise many distinct skills. The playworker is faced not only with the need to acquire, develop and apply these various skills, but also with reconciling the sometimes conflicting demands of the different elements of their work.

The balance between physical labour within the adventure playground site, and direct face-to-face work with playground users will be fluid, shifting in response, and subject to variations in the state of development of the site, seasonal constraints on work activities and the demand from playground users. Within the overall workload for the playground staff, a maximum commitment to the development and maintenance of the physical environment must be established. Each worker must be aware of the necessary inspection and maintenance procedures to be carried out, and their schedule (i.e. daily, weekly, monthly etc.). All other work commitments must be planned around this basic commitment, and all developments to the site must be planned in relation to their implications for future inspection and maintenance obligations. Intensive development of structures, play features and complex activity facilities may extend the possibilities of the play environment, but may also commit the playworkers to so intensive a 'caretaking' function that direct work with children and other playground users is only possible at the expense of inadequate attention to the safety of the site.

The playworkers must be aware of, and involved in, the way the environment is used. Most of the activity on the playground will be informal, chosen or initiated by the users and directed by them. The playworkers, while avoiding intervention in play activities which is directive, must identify patterns of activity which are potentially or actually hazardous.

Playworkers must also be aware of social pressures and groupings within the playground, and seek to minimize their effect on attitudes to activities. In particular the distortion of confidence produced by 'egging on' to feats of bravado or by the challenge of a 'dare' may result in children attempting hazardous activities beyond their physical ability, with a consequent high risk of injury.

In the early years of the adventure playground movement, many playgrounds were established and operated on minimal resources. Often such playgrounds were operated by a single worker. However, with increasing awareness of the health and safety requirements of these facilities and the importance and complexity of work with children and young people, the need for appropriate levels of staffing and resourcing has become accepted.

Where sufficient resources have been made available adventure playgrounds have continued, flourished, and provide a range of services to children and the communities which is often pioneering and unique, and always effective and value for money.

3.2 STAFFING LEVELS

It is strongly recommended that all adventure playgrounds have a minimum staffing level of three full-time playworkers or two full-time together with part-time workers equivalent to one full time worker. (Full-time in this context is defined as a minimum working week of 35 hours in a five-day week). A minimum complement of two permanent playworkers must be on site at all times during opening hours.

The actual staffing requirement may be greater and will be determined through an assessment of the resources required to meet the objectives set for the site. This will take into account the size of the site and its complexity, the range and level of activities, both on and off site, the level of demand from users, and any special factors, such as use by children with special needs. Staffing levels will also be determined with reference to risk assessment, and by the requirements of registration under the Children Act.

Permanent staff may be supported by other responsible adult(s) who can be casual, part-time employees or volunteers. Volunteers should be used to augment the permanent staff establishment never to replace it. Staff remain responsible for safety and operation on the site and this responsibility may not be delegated to volunteers. The use of volunteers carries with it additional, specific responsibilities for staff and managers, which are detailed in a separate section below.

Single staffing of any playground is not acceptable. It cannot be justified in the context of the safety requirements of these facilitates and the health and safety of the staff and the children. The management and staff should establish a rota which will meet the objectives set for the site, and take account of any Health and Safety and Children Act Registration requirements.

Adventure playgrounds catering for children with special needs require a much higher ratio of staff to children. Handicap or disability does not diminish the enjoyment of or the need for play, but more practical help, more individual attention and closer supervision may be necessary for the child to gain most benefit. HAPA – Play for Children with Disabilities and Special Needs, provide information and advice concerning staffing levels, see Appendix 4 for contact address.

Management and staff must encourage and support volunteers on the playground. The selection, training and placing of volunteers must be subject to agreed selection procedure.

3.3 LEGAL RIGHTS OF EMPLOYEES AND RESPONSIBILITIES OF EMPLOYERS

Employers must be aware of their legal duties and employees' rights. Employees rights fall into two categories; individual entitlements and rights acquired through trade union membership. This is a complex area which is thoroughly described in the booklet from National Council for Voluntary Organisations, *Voluntary Not Amateur*. A summary of the rights and responsibilities is provided below.

In summary: some employees' rights depend upon a worker's length of service and hours worked each week; however, all employees have the right:

- to work in a healthy, safe environment – Health and Safety at Work Act 1974 and Offices, Shops and Railway Premises Act 1963;
- to equal pay – Equal Pay Act 1970;
- not to be discriminated against on grounds of race (except where race is a

genuine occupational qualification for the job) – Race Relations Act 1976;
- not to be discriminated against on grounds of gender or marital status (except where gender is a genuine occupational qualification for the job) – Sex Discrimination Act 1975;
- not to disclose spent convictions or be penalised for the failure to disclose spent convictions (this right is not extended to playworkers because of their contact with children, see below) – Rehabilitation of Offenders Act 1974;
- to union membership and to take part in union activities;
- to hold a secret ballot in a workplace with a recognized union;
- not to belong to a trade union;
- not to be victimized, or unfairly dismissed on grounds of trade union membership or activities;
- to paid time off for ante natal care;
- to statutory sick pay (however there are variations in this);
- to compensation if they become ill or injured during the course of employment;
- a written statement of the terms of contract within 13 weeks of employment (for every person employed 16 or more hours a week).

In addition to the legal rights and responsibilities there is a wide range of additional measures which are good employment practice which are described fully in *Voluntary Not Amateur* (see above).

3.4 HEALTH AND SAFETY RESPONSIBILITIES TO STAFF

Employers have many responsibilities in respect of the health and safety of employees. For adventure playground staff these will include the following:

1. to undertake a risk assessment of all aspects of work by an experienced worker, and the undertaking of preventative and protective measures to combat the risks identified;
2. to ensure adequate working conditions, particularly in relation to such aspects as the structural condition and stability of the premises, means of access to and egress from premises, cleanliness, temperature, lighting, ventilation, overcrowding, noise, vibrations, dust, fumes etc.;
3. to provide welfare facilities for workers, including an adequate water supply, lavatories, washing facilities, first-aid arrangements, cloakroom accommodation, refreshment and rest facilities and drying facilities;
4. to make arrangements for securing the health of persons at work, including arrangements for medical examinations where appropriate, X-rays and health checks;
5. to provide protective clothing or equipment (including clothing affording protection against the weather), for example boots, wellingtons, warm windproof and water resistant jacket, goggles, industrial/rubber gloves, overalls, hard hats and safety harnesses, where necessary;
6. to provide such information and training in matters of health and safety as is required in order for work to be carried out to the proper standard.

3.5 RECRUITMENT AND SELECTION

Adventure playground staff should be recruited and selected within a framework of equality of opportunity. This will mean that applicants are selected on the basis of their relevant merits and abilities. This will entail:

- ensuring that all who should be are involved in the recruitment process from start to finish;
- the drawing-up of job descriptions which set out the aims and objectives of the job and the duties which are required;
- the drawing-up of person specifications, which identify the skills, qualities and experience needed for the job, as described in the job description;
- appropriate advertising, which ensures that people from all sectors of society are reached and those who may be suitable candidates encouraged to apply;
- a shortlisting process, which identifies suitable applicants by testing applications against set criteria;
- consistent interviewing for all candidates, based on agreed criteria drawn up from the job description and from the person specification and which tests for attitudes and practice on issues of race, gender and special needs.

More details of this process can be obtained from PLAYLINK.

The Children Act Guidance and Regulations require persons working with children under eight years old to be 'fit persons'. The fitness test involves consideration of the following.

1. *Previous experience*: for senior playworkers an absolute minimum of two years' working with children and an ability to demonstrate competence in a range of complex tasks; for other staff more limited experience of playwork can be satisfactory.
2. *Qualification and/or training*: there are now recognized National Vocational Qualifications in playwork at levels 2, 3 and 4. As these are relatively new qualifications and there may not be many candidates who have them, PLAYLINK recommends that the underpinning values and the competencies described in the NVQs be used as a guide in assessing whether a worker has appropriate skills for the job. Other recognized qualifications, e.g. in youth work or early years' education might be relevant in particular cases but their relevance would need to be demonstrated.
3. *Ability to provide warm and consistent care*: in the context of adventure playgrounds it is important that staff and volunteers are able to recognize the needs of, and respond appropriately to, each child.
4. *Knowledge and attitude to multi-cultural issues*: the recruitment panel should test candidates for attitudes and practice on issues of race, gender and special needs.
5. *Physical health*: a person specification will identify the physical capabilities required for the job concerned, and assurance must be sought from candidates as to their ability to fulfil duties capably.
6. *Mental stability, integrity and flexibility*: the recruitment process must test for

an understanding of what the job entails and the capability of doing it.

7. *Known involvement in criminal cases involving abuse of children*: this is discussed in detail in the following paragraphs.

There are also requirements under Children Act Guidance and Regulation concerning the vetting of the criminal records of persons living or working on premises where children under eight years are cared for.

Playwork is employment which is exempt from the provisions of the Rehabilitation of Offenders Act 1974. This means that candidates whose work will involve significant face-to-face contact with children must declare any current or spent convictions as part of the application process.

Playworkers who are recruited by a local authority to work (or volunteer) with children will automatically be vetted when they accept an offer of appointment. They must give written permission for this to be carried out.

For adventure playgrounds whose staff (or volunteers) are not appointed to local authorities and who do not have such checking carried out, the Guidance and Regulations do not insist on such vetting. They do, however (*Guidance and Regulations, Volume 8 – Private Fostering and Miscellaneous*), place strong emphasis on a suitable recruitment and selection process which includes the taking up of references and the thorough scrutiny of information in the application from.

PLAYLINK strongly advises that a police check be properly carried out on all persons working, or volunteering to work, on adventure playgrounds and who have or might have substantial access to children.

Where playgrounds do not have access to police vetting schemes, a rigorous recruitment process, management awareness and supervision can add safeguards. The Home Office in their code of practice for the safeguarding of the welfare of children and young people *Safe from Harm*, sets out 13 recommendations, which if followed enable voluntary organizations to minimize the risk of abuse. These include finding out whether an applicant has any conviction for criminal offenses against children, as one of a number of measures. Although at the time of publication of this book there are no additional vetting procedures available to voluntary organizations in order to meet this recommendation, a White Paper has been published which proposes measures to enable voluntary organizations to have access to police record checking procedures. It is envisaged that this will become law and take effect in 1998.

The Department of Health currently provides a consultancy service which relies on information provided by employers, and gives information on previous convictions, dismissals and resignations in certain circumstances of staff. This consultancy also holds a copy of 'List 99' which is operated by the Department of Education and Employment and lists the names of individuals who are barred by them from employment as teachers or workers with children and young people.

PLAYLINK emphasizes that, though many types of convictions should not automatically disqualify appointees, convictions for violence or abuse to children should.

Local authorities cannot compel police vetting as a condition of registration under the Children Act or grant aid.

3.6 CONTRACTS OF EMPLOYMENT

Every playworker must have a contract of employment. This legal agreement between the employee and the employer defines their respective rights and duties. The contract of employment, including procedures relating to conditions of service (such as sickness, discipline, grievance), must be agreed in writing within three months of the date of commencement of employment. The detail of the playworker's individual contract is of vital importance, since it stipulates hours of work, probationary period, the number of days to be worked in a week, holiday entitlement, and other general working conditions. Signed copies of the contract must be held by the playworker and the employer.

It is recommended that contracts for playworkers are supported by a detailed job description, which should specify general and specific responsibilities. Such a job description should refer to the safety policy in force for the playground as an integral part of the job and stipulate any particular requirements and responsibilities in respect of health and safety.

3.7 TRAINING

Training must be seen as part of the work and essential to the development of playworker and playground. Staff must be appropriately trained and, if appropriate, qualified to undertake the duties required of them.

Individual training needs of all staff must be assessed. This assessment will involve a comparison between the skills required to do the job and the skills currently held by the worker. This assessment must be recorded and reviewed regularly.

3.7.1 Health and safety training

In assessing the priority for training, specific training in skills and practice relevant to health and safety must be seen as a priority. The provision by employers of information, instruction, training and supervision to ensure safe practices at work are requirements of the Health and Safety at Work etc. Act 1974.

Employers must also:

- identify how the needs are to be met, creating suitable opportunities where they are not available;
- ensure that time for training is programmed into the playworkers' schedules;
- budget adequately for course fees and expenses involved in training staff;
- keep records of all training.

Playworkers have a responsibility to co-operate with management, by taking full advantage of health and safety training opportunities where they are arranged. They should also advise managers of any areas in which they consider they need training or information.

3.7.2 Induction training

It is recommended that all adventure playgrounds operate an induction training programme. Some form of familiarization with local conditions and systems is an essential and effective means of introducing the new worker or manager to the playground. This may include administrative and management systems, as well as the day-to-day operation of the playground. This programme should include a timetable for the completion of each section.

An integral part of the induction training programme must be familiarization with relevant standards of practice, with particular reference to the playground's own safety policy and procedures.

The induction process must be seen as bringing the workers or managers to a point where they can begin to fulfil their particular role and to identify particular needs for training. Where specific needs for training have been identified during the selection process, these must be formally identified and recorded as part of the training needs assessment, along with a timetable for suitable training. Trainees and students on placement will require appropriate induction training.

3.7.3 Other training

The introduction of the National Vocational Qualifications in Playwork based on the competence of performance at levels 2, 3 and 4 provides a framework for the assessment of training needs in a context which is broader than health and safety. Competence involves the assessment of a candidate's performance as measured against occupational standards for playwork. Although this system is based on competence in the workplace, often training will be required to augment an individual worker's knowledge and information.

Employers must provide opportunities for workers who wish to register and take part fully as candidates working towards the achievement of NVQs. In addition, workers must have opportunities to receive training in skills based on specialist areas, in order to develop the range of resources they are able to bring to playwork.

Employers must ensure that all training provided for staff is to appropriate standards and by those who are suitable to deliver the training.

The Children Act Guidance and Regulation Volume II requires that half of staff who work with the age group 5–7 years should be suitably qualified. See the Staffing section above.

3.8 VOLUNTEERS

Volunteers can be important assets to the work of the adventure playground. Many professional playworkers have begun, or developed their careers, as volunteers. They must be recruited and supported appropriately and instructed in all necessary procedures for their own safety and that of others. It is also important that they are not given sole or excessive responsibility, and are not required to fulfil either exclusively menial tasks or tasks in which they are not competent.

Moreover, working with volunteers involves responsibilities in respect of their recruitment, selection, worktasks and management.

3.8.1 Recruitment and selection

The recruitment and selection of volunteers to work with children alongside paid staff should mirror procedures for those staff (see above). These should include:

1. job descriptions to clarify tasks and lines of responsibility, define boundaries between paid staff and volunteers, define the knowledge needed, give status and limit opportunities for prejudice towards individuals; a job description should be written to assist, rather than exclude, potential volunteers;
2. appropriate advertising which ensures that people from all sectors of society are reached and that those who may be suitable candidates are encouraged to apply which emphasizes what volunteers can gain, how they can use their abilities and what is expected of them;
3. an interviewing and placement practice which responds to the needs of volunteers, provides accurate information, and challenges assumptions based on stereotyping.

The criminal records of volunteers who will be working with, or have substantial access to, children will need to be checked, in the same way as for paid staff, see above.

3.8.2 Training

Many volunteers will want to use their volunteering as a means to personal and professional development. They should be encouraged to take up training. Training of volunteers should aim to help them to be more effective in their voluntary work. The individual training needs of all volunteers must be assessed. The assessment should be recorded and should be reviewed periodically.

Training for volunteers must always include induction and health and safety training. Skills training and professional training may also be appropriate.

Where training needs are to be met through formal training it is important to ensure such training is arranged in a way that it is accessible.

3.8.3 Liability and insurance

The liability position of volunteers is not the same as for paid staff. Under English law the employer is, in almost all circumstances, liable for injuries caused by their employees to their colleagues. This is called vicarious liability. Vicarious liability is insured against through the employers' liability insurance arrangements.

The principle of vicarious liability does not extend to volunteers, and therefore they are not normally covered by the employers' liability insurance. It is important for organizations who rely on volunteers to ensure that they take out, or extend, this insurance in order to cover liability for injury to colleagues due to the negligence of a volunteer.

Liability for injury, loss or damage to persons other than employees caused by the negligence of a volunteer must be insured against through the public liability insurance arrangements made by the adventure playground. The insurers must be made aware that volunteers are used as part of the operation of the facility. Some insurers have experience of insuring adventure playgrounds and it is worth seeking them out. Public liability insurance for a claim of up to £2 000 000 is recommended.

Other insurances relevant to the use of volunteers include:

- personal accident – to compensate for accidents to volunteers where no liability (fault) is involved;
- property insurance – if a volunteer suffers loss of their own personal property while volunteering;

The actual insurance position of volunteers must be made known to them before they accept any offer of placement that may be made to them. They may wish to insure themselves.

3.9 FURTHER READING

The following are the main source documents for this chapter:

Advice and Development of Volunteering and Neighbourhood Care in London (Advance) (1990) *Equal Opportunities and Volunteering. A Guide to Good Practice*, Advance.

Children's Legal Centre (1992) *Working with Young People: Legal Responsibility and Liability*.

Children's Legal Centre Information sheet *Children, Safety and the Law*, undated.

HAPA (1994) Insurance for Children's Play and Legislation and the Playground, *HAPA Journal*, **13**, pp. 18, 19.

Kids' Clubs Network's Police Check Scheme, undated.

Kids' Clubs Network, *Keeping safe*, undated.

London Voluntary Service Council (1994) *Voluntary But Not Amateur*.

Pedlar, P. (ed.) (1983) *Insurance Protection: A Guide for Voluntary Organisations and for Voluntary Workers*, NCVO.

PLAYLINK, *Open Access Play and the Children Act*, undated.
PLAYLINK, (1988) *Recruitment of Staff*.

Chapter 4

Adventure playground site and building

4.1 SITE SELECTION

The siting of an adventure playground has implications for future health and safety practice. It is not within the scope of this publication to consider all the factors which should be taken into account, but the following points are critical:

1. The site must be accessible to the proposed catchment area, as an attractive facility will draw children. Care must be taken with the routes to and from the site to ensure that they are safe for children to use.
2. Direct access from the playground onto such hazardous sites as public car parks, main roads, railways, canals and rivers must be avoided.
3. Proximity to industrial sites must be avoided.
4. Attention must be paid to establishing pedestrian crossings, prohibited parking areas and signs warning drivers of the presence of children.
5. Access for emergency services, cars, minibuses and lorries must be possible, with loading areas well lit and with sufficient space for users with wheelchairs to embark and disembark from their transport.
6. Access for pushchairs, wheelchairs, prams and wheeled toys will be needed.
7. Sites close to overhead power or telephone lines must be avoided.
8. Opportunities for the informal supervision of the access routes to and off the site must be considered.
9. All relevant public utilities (e.g. water, gas, electricity and telephone) must be consulted to ensure that the site is clear of underground services which would be likely to restrict construction works or subsequent playground activities and developments.
10. Mains services must be within reasonable distance of the site to allow installation at minimum cost and ensure adequate capacity for the needs of the adventure playground.

11. Natural drainage of the site, and the possibility of improvement if it is not adequate, must be considered.
12. The site must be stable and free of contamination, particularly demolition sites, where old basements, drains and other voids must be correctly backfilled and the surface made good.
13. An investigation must be undertaken, to establish whether any toxic waste has been dumped on the site, and for the presence of radon.
14. Landfill sites must be avoided.
15. The site, buildings and other facilities must be large enough to accommodate all the children who are likely to use them.

Adventure playgrounds must provide opportunities for children of all abilities to mix together. It is therefore important that the planning, design (and management) processes ensure that the playground is accessible to children with disabilities and special needs.

Specialist knowledge and advice should be sought at the earliest stage of the planning process, including contacting the local authority planning department.

4.2 SITE LAYOUT

Following the selection of the site it is necessary to prepare a detailed plan of the proposed permanent features of the playground. The location of major features, such as the building, will need to take account of the location of the mains services, access and other planning requirements. It is, therefore, essential to undertake (in conjunction with the statutory authorities) a survey of the site and prepare a site services drawing.

The layout will be considerably influenced by these factors and the importance of early detailed planning cannot be over stressed. The advice of end-users, planners, architects and surveyors must be sought in order to avoid errors in layout, design and construction. It will be necessary to keep in mind that the site and buildings are to be child-centred, and that the site must allow for variety and change in the physical environment.

Buildings must be designed in a way which makes them accessible to people with disabilities. BS 5810 is a Code of Practice of basic architectural provisions that must be incorporated into new buildings to make them convenient for the use of wheelchair users and those with hearing and/or sight impairments. Its recommendations may also be used as guidelines for the adaption of existing buildings.

Locating the play building on the edge of the site has many advantages because it allows direct access to the building when it is not appropriate for the outside area to be used.

4.3 SITE SECURITY, FENCING AND ACCESS

Adventure playgrounds, as defined within this document, must always be securely enclosed. Play equipment in an area which is not enclosed will have to conform in all respects to the recommendations of BS 5696: Play Equipment Intended for Permanent Installation Outdoors and where that is not achievable to DIN 7926.

Perimeter fencing must be safe, secure and complete to a minimum height of 2 m. This can be timber stockade, brick, weldmesh, combination brick and weldmesh, concrete panel etc. While children should not be able to climb the fencing easily, it must be capable of supporting their weight. It must be securely fixed to the ground, to prevent undermining. In some locations it is desirable to provide fencing which allows for informal supervision from outside the site without destroying the sense of privacy for playground users. Perimeter fencing must be as resistant to vandalism as possible. Strong and solid fences or walls provide for maximum security. Barbed wire and spiked fencing must never be used as any part of the boundary or elsewhere on the site.

The provision of secure fencing is necessary as part of the discharge of duties under the Occupiers' Liability Acts of 1957 and 1984, which give a duty of care to trespassers who may be at risk of injury. In order to maintain security:

• fencing must be solidly constructed and regularly maintained;
• materials must not be stacked against fencing;
• structures and landscape trees must be sited well clear;
• gates must be of heavy-duty construction with adequate hinges and fittings, and securely locked when the site is unsupervised.

The provision of a 1 m soft landscaping strip outside the fence, planted with shrubs may reduce damage to the fence and reduce its visual impact.

The use of notices giving explicit warnings concerning the hazards to trespassers will assist in the discharge of this duty. They must be in plain language, and pictorials may be appropriate. If necessary these must be in more than one language.

Before closing the playground, staff must ensure that all children are off site and that all reasonably practicable steps have been taken to reduce risks to trespassers, by:

• immobilizing or dismantling any moving equipment, such as swings or an aerial runway;
• ensuring that stacked materials do not represent a hazard;
• removing ladders and other items of loose equipment to a secure store;
• removing all tools and play equipment to storage;
• emptying waste containers disposing of waste safely and securely. Paladin-type bins must be securely locked to a permanent feature to prevent them from being moved or overturned;
• securely boarding over any holes or excavations;
• locking all buildings or stores;
• fully extinguishing any fires;
• turning services off at the mains where practicable.

A closing down checklist had been provided at Appendix 5.

Vehicular and pedestrian accesses should be separated. To counteract the effects of intensive use, the access paths should be appropriately surfaced. Access should be from more than one part of the site if possible, to provide an escape route in the event of fire.

Vehicular access should be through double gates with a minimum clear opening of 3 m, which must allow easy access for emergency services. Visitors must not be allowed to drive into and park on site. The vehicular access should be sited to allow all deliveries to be made directly into well-defined storage areas. Vehicular access gates should be kept locked until there is a need for their use. When vehicles enter the site, children must be kept well clear. Deliveries of materials should, where possible, be undertaken when the playground is not in use.

Pedestrian access gates must have a minimum clear opening of 1.2 m. If pedestrian access is directly onto the pavement, there should be a pedestrian barrier between the playground and the road. Consideration should be given to the provision of warning signs for motorists, restrictions on parking in the vicinity of the entrance, and the provision of a pedestrian crossing.

One or more large and easily readable permanent signs must be provided stating:

- the full name, phone number and address of the playground;
- the opening times, both in school term time and during school holidays;
- a contact address for the management body, with an emergency contact procedure;
- that entry to the site is not permitted outside the displayed opening hours or when the gates are locked, and that children may be at risk if they gain/force entry outside these times.

Provision must be made for temporary notices on the main notice-board to inform of closure due to staff holidays, illness or some other unforeseen circumstances.

Such notices must be thought of as part of, not instead of, health and safety measures. Under the Unfair Contract Terms Act 1971, they cannot exclude or restrict liability for death or personal injury resulting from any negligence on behalf of the operators of the site, but if the notice and the terms on them are reasonable then they may exclude or restrict liability for loss or damage. Where children are concerned there are limits to what they can be expected to understand from a written notice. For example, clock faces, with opening and closing times, can be provided.

All notices must be welcoming and should be written in languages in common use in the community in which the adventure playground operates.

The police and fire service should be informed in writing of the times when the playground is normally closed and the names and addresses of the playground keyholders.

More than one keyholder should be provided. The safety of any keyholder who is required to attend the site when it is closed must be assured, through safe travel arrangements and close liaison with police.

Members of the management body, or other responsible persons, living close to the playground, should be provided with duplicate keys to the site and the building in case of an emergency requiring access to the site outside opening hours.

4.4 SITE SAFETY

The outside area (site) of an adventure playground is a place of work under the terms of the Health and Safety at Work etc. Act 1974 and its Regulations in the same way that the indoor areas are, and therefore there are general and specific health and safety requirements which relate to it.

Sections 2 and 3 of the Health and Safety at Work etc. Act 1974 place duties on employers, as follows:

> to ensure, so far as is reasonably practicable, the health, safety and welfare of employees, and to conduct his undertaking in such a way as to ensure, as far as is reasonably practicable, that persons not in his employment who may be affected thereby are not thereby exposed to risks to their health and safety.

The Management of Health and Safety at Work Regulations 1992 requires employers to assess risks which arise from their undertakings, to prioritize the measures that need to be taken in order to control the risks identified and to comply with any relevant legal requirements.

Two legal requirements which concern the health and safety of an adventure playground site arise from the provisions of the Workplace (Health, Safety and Welfare) Regulations 1992. These affect all workplaces which were used for the first time after 31 December 1992, and existing workplaces on 1 January 1996, and are concerned with protecting the health and safety of everyone in the workplace. A written poster on these is available from the Health and Safety Executive.

The first requires that floors and traffic routes be provided and maintained in such a way as to remove the risk of slipping, tripping or falling caused by uneven, slippery or damaged surfaces. This means that on the adventure playground site:

- surfaces where lifting and carrying of objects is required must be suitable for the purpose;
- there should not be trip points, uneven and/or slippery pathways.

The second requires that all reasonable steps be taken, through the provision of facilities, to prevent falls by a person or a person being struck by falling items. This means that on the adventure playground site:

- storage of materials must be done in such a way as to avoid items falling;
- items must be stored heavy low, light high where this is practicable;

- items must not be stored above a height of 2 m;
- guardrails must be provided where a fall of 2 m or more is possible;
- the need for people to climb on top of vehicles or their loads must be avoided as far as possible, and where it is unavoidable, effective measures must be taken to prevent falls.

4.5 BUILDINGS

Adventure playground buildings must be safe for those who work in them and those who use them. Where the users include children under eight years of age, there are specific standards concerning the quality of the building and its facilities, as follows.

4.5.1 Construction and layout

Buildings must be soundly constructed and suitable for the purpose(s) for which they are to be used. The Construction (Design and Management) Regulations 1994 require that those for whom a construction project is carried out (clients) must be reasonably satisfied that they only use competent people as supervisors, designers and principal contractors, and that sufficient resources, including time, have been allocated to enable the project to be carried out in compliance with health and safety law. This applies to construction projects which will last longer than 30 days or 500 person days. More details of these Regulations are provided in Part Two of this publication.

It is good practice to ensure that no elements of building construction and layout prevent use by children with disabilities.

Suitability of premises for the provision of day care facilities, which include adventure playgrounds, under the Children Act will require:

- fixtures and fittings to be appropriate and safe;
- access to, and standard of, kitchen or space for the provision of meals and snacks if required;
- suitable, adequate and rehearsed fire fighting procedures (fire drills at least twice a year, see Chapter 9) suitable fire-fighting equipment and adequate exits;
- appropriate heating, lighting and ventilation systems;
- sufficient toilets and washing facilities;
- that rest area and access to office/administrative base and (emergency) telephone be provided;
- that premises be suitably clean and a reliable system for cleaning;
- that premises be at a suitable location within the children's community;
- keyholding arrangements, including planning for emergency arrangements.

Any changes to the design and layout of the building, or major changes to the use or uses within the building, must be discussed with the fire service in order to ensure that suitable fire safety measures are in place.

4.5.2 Entrances and exits

Entrances and exits from the building must be adequate in size, number and location. All exits must be clearly marked or illuminated, free from obstruction and unlocked at all times when the building is in use. Doors, hinges and frames must be sound and secure and must open outward easily in such a way as not to give rise to an unacceptable risk. Door fittings must be properly maintained. If possible the hinge side of doors must be guarded to prevent injury to fingers. In some circumstances it may be appropriate for doors to be fixed open.

Issues concerning access which will be considered in the registration of premises under Children Act Guidance and Regulations are:

- suitable access to an outside play area;
- controllable access to other areas in a shared or multi-use building;
- proper access for children with disabilities;
- an identifiable and welcoming entrance area;
- precautions to prevent children from exiting onto the road.

Where there is direct access from site onto a road or car park, pedestrian barriers, warning signs and traffic calming measures must be provided.

Consideration of fire precautions will also be included in the registration of premises. This may be done by the Children Act registration team or by fire service staff. In either event it is likely that the requirements and recommendations of the Fire Precautions Act of 1971 as amended by the Fire Safety and Safety of Places of Sport Act 1987 will be used as a basis for the assessment, although these principally apply to premises licensed for public entertainment.

This legislation requires the following of all premises to ensure satisfactory means of escape:

- all doors must work and be free of obstruction;
- all fire doors must be indicated;
- sliding doors must be marked with direction of movement;
- all doors must be marked and unlocked and gangways and escape routes must be clear when the premises are in use.

Appropriate smoke alarms must be fitted in passageways and strategically placed throughout the building. The advice of a fire officer must be sought on the location of these.

More information about the impact of the Fire Safety Acts on adventure playgrounds is given in Chapter 9.

It is helpful to the flexible use of the adventure playground if the access to the toilets is possible from outside the building at times when the rest of the building is closed.

All safety signs must conform to the requirements of BS 5378. The manufacturers of safety signs provide excellent information on the types and uses of safety signs.

4.5.3 Floors, passages, stairs

Floors, passages and stairs must be of sound construction and properly maintained. They must be kept free from obstruction and from anything likely to cause people to slip, for example, mud or liquid spillage.

Floors must be free from holes, slopes and uneven or slippery surfaces.

Staircases must not be pitched too steeply. Forty degrees is the optimum, but the shallower the pitch the safer and more convenient for use by children. All staircases in buildings, or in use as exits from buildings, must have a substantial handrail. Consideration should be given to the provision of a second, lower hand-rail for use by smaller children. Open sides of staircases must be filled in with solid panels or vertical rails with gaps between no wider than 100 mm.

4.5.4 Walls and roofs

Walls must be sound, strong and free from damp or condensation, cracks or any other evidence of fault or damage. The Local Authority's Building Inspectors must be consulted before any structural alteration is undertaken. The installation or modification of doors or windows must be carried out in accordance with Building Regulations, or structural strength may be affected.

Except where specifically engineered as part of the play environment roofs must not be easily accessible to children. Nonetheless inevitably a child will eventually climb up so they must be so designed and constructed as to be able to support the weight of children who gain access onto them. Practical measures may be taken to reduce access to roofs, for example, the use of non-drying 'anti-vandal' paint. Where this is used a warning notice must be posted.

Roofs must be adequately drained, with guttering and down-pipes into gulleys. All rain water pipes must be cleared regularly and repaired promptly when damaged, to reduce the risk of problems caused by damp. Downpipes should be protected from accidental damage. Gulleys should be suitably covered, to prevent their becoming blocked by debris. Plastic guttering and downpipes may not be suitable for use on an adventure playground.

4.5.5 Glass

Registration of premises under the Children Act will include an assessment of the provision of glass to appropriate safety standards.

Windows must be sound and secure. If these are below waist height (a sill level of 800 mm or less) they should have safety material installed, or be adequately protected against breakage. This could be a polycarbonate, or georgian, toughened or laminated glass. The choice of the appropriate material for internal glass may require consideration of their fire resistance, and must be checked with the fire service. If windows are cracked or broken they must be replaced immediately and the old glazing materials disposed of safely.

Windows and skylights should be capable of being opened and of being cleaned without exposing the operator to risk. They should be kept clean on both sides. Windows must not open outwards directly onto pathways or into areas where persons are likely to collide with them. They must be capable of being secured in the open position to prevent accidental slamming shut. Window sills must be kept clear.

The Workplace (Health and Safety) Regulations 1992. require glass found below shoulder height in doors to be of safety material or be adequately protected against breakage. However it is recommended that all glass indoors is of a safety type. The choice of appropriate glass will almost certainly include consideration of fire resistance and must be made in consultation with the fire service.

Glass or transparent materials in doors and/or windows are especially susceptible to breakage, i.e. in doors and gates below shoulder height and in windows, walls and partitions at waist level and below, which should be marked where necessary to make them apparent.

4.5.6 Storage

The provision of adequate and appropriate storage is not only good practice, it is also a requirement of the Health and Safety at Work etc. Act 1974 and its Regulations, and of the Children Act, through its Guidance Regulations, Volume ll. Registration of premises under the Children Act will include an assessment of adequate storage.

All storage areas must be dry, with shelving, racks and cabinets strong enough for their intended use. External access may be advantageous in some cases. All stores should be lockable with security levels appropriate to the nature of the materials stored therein. Stores should be kept tidy, be checked regularly, and no materials should overlap or protrude from shelf edges. An up-to-date inventory of all contents must be kept.

Toxic, flammable or corrosive materials must be kept to an absolute minimum, and must be stored securely and in accordance with manufacturers' instructions. Some materials must not be stored together, for example corrosive substances and flammable materials. The appropriate storage of materials should be identified as a result of the risk assessment undertaken as part of the Control of Substances Hazardous to Health (COSHH) Regulations 1994. Product data sheets obtained from the manufacturer give information on the storage of materials.

As a rule of thumb, any materials labelled 'Keep out of the reach of children' should be treated as hazardous.

Areas where hazardous materials are stored must be marked with warning signs. Lists of the hazardous materials stored must be posted at the place of storage.

Storage should be provided for all personal protective equipment and for workplace clothing, or changes of clothing, including children's clothing.

The construction and use of storage facilities must prevent the risk of falls by a person, or a person being struck by falling items. This may necessitate provision of appropriate steps, ladders or hop-ups.

Particular care must be paid to ensure that sharp tools, power tools and heavy objects are stored correctly and that items cannot be accidentally dislodged from racks, shelves etc.

Areas where electricity or gas services are housed must not be used for storage, unless the consumer units/meters are properly enclosed.

The safety and security of storage areas should take account of the needs of those who are not easily able to distinguish between different liquids and powders.

Details concerning the storage of foodstuffs is provided in Chapter 8.

4.5.7 Heating and ventilation

The temperature, ventilation and lighting of an indoor workplace is subject to the requirements of the Health and Safety at Work etc. Act 1974 and its Workplace (Health and Safety and Welfare) Regulations 1992. Registration under the Children Act 1989 will involve an assessment of heating requirements ventilation and lighting. The Offices, Shops and Railway Premises Act 1963 makes requirements in respect of heating levels in administrative areas.

The Health and Safety Regulations (above) are concerned with the employer's duty to provide heating levels which give reasonable comfort without the need for special clothing. This is normally at least 16 °C. However, for an adventure playground building where children are present, a temperature of 20 °C is more appropriate. One hour after opening the temperature inside the building must be not less than 15.5 °C. Where children under five years old use the building, higher temperatures will be required.

Heating and ventilation systems must be adequate to enable comfortable temperatures to be maintained. Heating appliances must be kept in proper working order and suitably protected to keep children safe. They must be regularly serviced by a qualified and competent person. Portable heaters such as paraffin or calor-gas stoves must not be used.

There should be adequate circulation of fresh air throughout the building.

Where low level heating units are used they should be secured to walls and adequately guarded.

4.5.8 Lighting

Lighting must be sufficient and suitable. Activity areas must be particularly well lit, as must passage-ways and aisles and the top and bottom of stairs. The use of natural daylight must be maximized wherever possible.

If fluorescent lighting is used the tubes and diffusers should be cleaned at least once a month to maintain effectiveness. Tubes and starters should be replaced at regular intervals in accordance with manufacturer's instructions. Diffusers should be fitted to reduce hazards from flicker which starts long before it becomes obvious and can cause epileptic seizures. Diffusers also provide protection from objects which may strike the tubes.

A supply of spare tubes and bulbs should be kept, in order to allow prompt replacement of defective lamps. A number of torches in good working order should be available in the event of power failure.

Where emergency lighting is provided it must run off a source of supply independent of the mains. It must be checked at regular intervals.

4.5.9 Fire precautions

Consideration of fire precautions will be included in the registration of premises under the Children Act. This may be done by the Children Act registration team, or by fire service staff. In either event it is likely that the requirements and recommendations of the Fire Precautions Act of 1971, as amended by the Fire Safety and Safety of Places of Sport Act 1987, will be used as a basis for the assessment, although these principally apply to premises licensed for public entertainment.

Fire Precautions must be adequate and advice must be sought from the local fire officer. Such advice will include both preventative measures and how to deal with emergencies. Preventative measures are discussed below, and fire precautions as a part of emergency procedures are set out in Chapter 9.

The design and construction of adventure playground buildings will have included provisions for preventing or limiting the effect of fires, through the application of building regulations. Where any alterations or additions to premises are considered, further advice must be taken to ensure the continuing effectiveness of these provisions.

A satisfactory means of escape must be provided. This entails ensuring that:

- all doors work;
- all fire doors are indicated;
- sliding doors are marked with the direction of movement;
- all doors are marked and unlocked and gangways and escape routes clear, whenever the buildings are in use.

Any notices must be capable of being understood by children, and if necessary be in more than one language.

Each building must have appropriate means of fighting fire, checked regularly in accordance with manufacturers' instructions and readily available for use. Staff must be trained in the use of such equipment. On an adventure playground these will normally include fire extinguishers and fire blankets.

Extinguishers must be of a type suitable for the class of fire. The following types of fire extinguisher are available:

- water – colour red: suitable for ordinary combustible fires, for example wood and paper, but are not suitable for flammable liquids;
- foam – colour cream: suitable for small liquid spill fires;
- dry powder – colour blue: suitable for flammable liquid fires;
- carbon dioxide – black: suitable for fires involving electrical equipment.

The green vaporizing liquid (Bcf) extinguishers are no longer recommended, as they are environmentally damaging and give off toxic fumes.

The types of extinguishers provided must be suitable for the risks involved. They must be fully and be clearly marked, with an information sign provided alongside. They must be adequately maintained and a record must be kept of all maintenance, inspections and tests. Staff must be trained in the use of fire extinguishers.

Fire blankets must be located where the risk of fire to an individual is greatest. This is normally in the kitchen. The instructions concerning location, use and maintenance of fire blankets provided by the manufacturer must be followed.

It is recommended that adventure playgrounds operate a non-smoking policy. Children Act registration requires that there be no smoking in front of children. Rubbish must never be allowed to accumulate, and must be removed every night.

Furniture which contains foam must not be used unless it has been tested as prescribed, and labelled in accordance with the Furniture and Furnishings (Fire) (Safety) Regulations, 1988. Foam furniture must have fillings which comply with the Regulations and covers which pass the match test set out in BS 5852 1982, Part 2, or have a suitable interliner. All new furniture should carry a permanent label which gives information about the tests which the furniture has passed, and contain the words 'CARELESSNESS CAUSES FIRES'. The labels are usually sewn into a seam beneath cushions or underneath the frame. Adventure playgrounds should not accept donated second-hand furniture, unless it has one of these labels.

These Regulations also apply to all post 1950 second-hand furniture sold in the course of trade, which should be sold with a triangular display label with a red border and the word 'CAUTION' in white. They will not necessarily have an information label and must not be purchased if they do not.

Curtains must be non-flammable.

Electrical appliances (except those designed to be left on, check with the manufacturer) must be switched off and unplugged as part of the closing-down procedure.

Soft playrooms and foam-filled play equipment and matting are often provided at adventure playgrounds. Foam used in these items should be Combustion Modified High Resilience and the PVC which covers the foam should have fire retardant properties, and certification from an accredited test house should be obtained by purchasers. *Safety in Indoor Adventure Play Areas* gives more details of this subject. (See Further reading, below.)

This equipment must be stored in such a way that, if set alight, the consequences are as far as possible minimized. This could include provision of separate storage area, with fire check doors which are kept closed. The storage of this equipment must be considered as part of the fire prevention measures.

The provision of smoke detectors is recommended. Where they are used they must be maintained according to the manufacturers' instructions. This includes testing at least once a month and cleaning and replacing batteries once a year. In dusty atmospheres, cleaning may need to be more frequent.

Staff must be given appropriate training in fire safety.

4.5.10 First aid facilities

First aid facilities must be provided. A separate area of the building should be set aside for first-aid treatment. There must be adequate space (warm and ventilated) for the injured child or adult to lie down and for the first-aider to treat the injury. The area set aside must be easily accessible for transportation of the injured person by stretcher or wheelchair if necessary.

The area should be provided in a way that ensures that it can be maintained in a hygienic condition. This could include the provision of:

• heavy-duty washable flooring surfaces wherever practical, porous floors must be sealed and the sealing process periodically repeated;
• walls which are easy to clean;
• surfaces which are sealed;
• a refuse bin with a lid;
• hand-washing facilities;
• storage for first-aid equipment and report books;
• a rest and recuperation facility.

The adventure playground's first-aid equipment must be kept in a secure container, marked with a white cross on a green background. The first-aid area must be cleaned daily.

Chapter 9 provides information on first-aid treatment.

4.5.11 Toilets

The provision of suitable and sufficient sanitary facilities for staff is a requirement of the Workplace (Health, Safety and Welfare) Regulations 1992. The suitability and sufficiency of these facilities for children will be considered as part of Children Act registration.

Wherever possible separate toilet facilities must be provided for males and females, to include those suitable for people with disabilities. Sufficient toilets must be provided so that children do not have to queue to use them. Separate facilities for staff must be provided.

Washbasins must have hot and cold running water and soap available. The temperature of the water must be properly regulated and not exceed 30 °C. Paper towels or electric warm-air hand-driers are recommended. Roller towels are not recommended as they have been known to be hazardous to younger children.

Facilities for the disposal of sanitary towels must be available.

4.5.12 Kitchens

The health and hygiene of kitchens (and food handling), even where these are a play activity, are subject to the requirements of the Health and Safety at Work etc. Act 1974 and its Workplace (Health and Safety and Welfare) Regulations 1992, and the Food Safety Act 1990 and its Food Hygiene Regulations 1991.

Registration under the Children Act 1989 will involve an assessment of standards of areas for food preparation.

The Health and Safety at Work etc. Act and its regulations (see above) impose responsibility on employers for the general cleanliness of premises, and for the maintenance of suitable hygiene standards for facilities where employees prepare or eat their own food.

The Food Safety Act 1990 sets standards for the safe handling and preparation of food. These include consideration of standards of equipment, storage and general cleanliness, e.g.:

- equipment – must be appropriate for the job required and in good condition;
- storage – the safe storage of food requires refrigeration below 5 °C. Frozen foods should be stored at minus 10 °C. Cooked and raw foods should be stored separately;
- cleanliness – all surfaces are to be kept clean, using a disinfectant daily.

The kitchen should be designed and fitted so as to facilitate cleaning of all surfaces, which should be of impervious materials, such as stainless steel or plastic laminate to facilitate cleaning and avoid contamination. Cookers and other equipment should be moveable, not built in, to enable cleaning behind them.

A fire blanket must be provided in the kitchen. Smoke alarms are not practical in a kitchen area, but heat-activated warning alarms are advisable.

The preparation of food at camps or barbecues etc. should follow these standards as far as is reasonable. More details on food hygiene can be obtained from the local authority environmental health department.

4.5.13 Welfare facilities

Provision for staff rest must be made and include proper and sufficient seating facilities, drinking water and facilities for drying clothes.

4.5.14 Providing for children with disabilities

All buildings should aim to comply, and all new buildings should comply with the British Standard Code of Practice for *Access for the Disabled to Buildings*, BS 5810.

4.5.15 Electricity

Electricity can be extremely hazardous. The health and safety aspects of the use of electricity have been broken down in this publication into two parts, installations and equipment. The safety of equipment is covered in Chapter 7, and the following section deals with the safety of fixed installations.

The Workplace (Health, Safety and Welfare) Regulations 1992 and the Electricity at Work Regulations 1989 include regulations concerning the safety of electrical systems.

Fixed electrical installations such as ring mains must be installed in accordance with IEE (Institution of Electrical Engineers) Regulations. They must be inspected every five years, and an inspection/test certificate obtained. If the installation is likely to suffer damage or abuse it must be inspected more frequently.

When installations are modified they must be checked by someone competent to do so before they are brought into use.

All electrical sockets, fittings etc. must be chosen according to the use to which they will be put. Routine inspection must identify damage and trigger immediate remedial action.

The positioning of sockets must be carefully considered. It may be appropriate to locate sockets out of the reach of younger children. Sockets must be away from traffic flows and protected from damage by furniture etc. Leads to appliances must not cross traffic flows and must not represent a hazard to the appliance user or others, by snagging or tripping. Where patterns of use change, or unforeseen demand occurs, new permanent sockets must be installed. Sockets must be fitted with safety caps when not in use.

The use of adaptors or extension leads is to be avoided. However if such use is unavoidable, for example to serve a hand-held power tool, flexible cables must be selected, maintained and used so that there is adequate protection against foreseeable mechanical damage. This includes fully uncoiling the flex when in use to prevent damage by heating. Extension cables used outside the building must be provided with a transformer reducing the voltage to 110 volts.

The installation of the lighting system must include consideration of the siting of switches with regard to the use of the building and access by children. It is useful to have one master switchboard which can be secured against interference by children. All switches must be marked to indicate which fittings they control. It must be possible to complete the closing procedures with adequate lighting.

Lighting fixtures must be securely fastened and protected from accidental damage where possible. Adequate stocks of spare bulbs/tubes must be maintained. When replacing bulbs/tubes the fitting must be isolated, and the work must be performed from a stable position, preferably a set of steps.

The connection to the mains supply, company fuse and isolator, and the consumer fuse and distributor boards must be sited so that they are not susceptible to casual interference by children, without impeding access to the isolator and individual circuit fuses by staff.

The system must be so designed that the failure of any individual circuit will cause minimum inconvenience – for example, if the lighting is split into two circuits, the failure of one will not plunge the building into total darkness. Each circuit must be clearly marked on the distributor board.

It is usual to fit Earth Leakage Circuit Breakers (ELCBs) to each circuit on the board, because of their speed of response and ease in resetting. Where wire-carrier fuses are fitted, an adequate supply of each capacity of fuse-wire must be available at the distributor board.

If ELCBs are not fitted, Residual Current Devices (RCDs) must be used for portable equipment and must be used for equipment used outdoors.

They are essential where electrically operated hand-held portable equipment is used and must be used where the electrical equipment is used outdoors as back-up protection against electrical shock. The use of low-voltage equipment, with current correspondingly transformed, is required wherever equipment is used outdoors. Health and Safety Executive Guidance Note PM 32 gives full details of the safe use of portable electrical equipment.

If installed RCDs must be tested frequently by means of the test button on the unit and, when the installation is routinely inspected (see above), the tripping current and timing of RCDs must be checked.

A torch and screwdrivers of the appropriate size must also be available. Improvised repairs to fuses must never be carried out, and only fuse-wire of the appropriate capacity for the particular circuit must be used. (In general, maximum values used are: lighting – 5 amps; ring circuits (plugs) – 15 amps; and cookers – 30 amps.)

Where electrical installations, plugs and sockets are provided outdoors they require specialist wiring and fittings.

In Guidance Note GS 50 (*Electrical Safety in Places of Entertainment*) published by the Health and Safety Executive in May 1991, there are some useful guidelines on the electrical safety of lighting and sound equipment. This is available from the Health and Safety Executive.

These have been drawn up in the context of a number of accidents in which entertainers have received electric shocks from their equipment, some of which have been fatal.

They are of relevance to those adventure playgrounds which provide for entertainments using these media, and it is recommended that GS 50 is obtained as a reference document.

4.5.16 Gas

The Gas Safety (Installation and Use) Regulations 1994 relate to most gas installations including Liquid Petroleum Gas (LPG) installations. These require that:

- work in relation to gas fittings must be carried out by a competent person;
- employers must take reasonable steps to ensure that anyone employed to do work in connection with gas fittings is registered with CORGI (in effect these two requirements mean that employers have an absolute duty to ensure that anybody employed to carry out work on a gas installation must be registered with CORGI, and to take reasonable steps to check on CORGI registration);
- employers must ensure that gas appliances or installation pipework installed at a place of work are maintained in a safe condition;
- all gas appliances (with the exception of portable mobile appliances supplied with gas from a cylinder) must be checked at least every 12 months by a CORGI registered person.

All gas appliances and supply equipment must be installed with consideration of the use of the building by children. Supply pipes must be adequately protected against damage, and all appliances protected against interference or accidental contact with hot surfaces.

Meters and isolating valves must not be susceptible to interference by children, but must be accessible to staff or others in case of emergency.

All appliances must be fitted with automatic safety cut-outs so that if the pilot light is not lit, or fails to re-ignite if blown out, the appliance is isolated from the supply. Gas cookers, where possible, must have automatic ignition by pilot light or piezo crystal to prevent accidental build-up of gas (and to minimize use of matches etc.).

If there is any suspicion of a gas leak the building must be evacuated, the gas must be turned off at the mains and doors and windows opened to disperse the gas. There must be no smoking or naked flame of any description and the use of electrical equipment must be avoided. In the event of any leakage, or suspected leakage, British Gas must be informed immediately and no appliance used until inspected and authorized for use by their representative.

The use of LPG in canister form must be avoided at an adventure playground. Canisters may be required for camping etc and these must be returned to the supplier as soon as they are no longer used.

4.5.17 Water and drainage systems

Water, plumbing and drainage systems, though less intrinsically hazardous than other main services, must be regularly maintained.

Water pipes, where possible, must be sited to prevent possible damage or interference. It is advisable to include a stop-valve for each separate installation, to facilitate repair and maintenance.

All water installations must conform to local regulations and be inspected by the water company before connection to the mains. Cold-water storage tanks, where fitted, must be covered. All piping and tanks must be well insulated against frost damage. All tanks, basins etc. must have adequate overflow and outlet pipes. Taps and valves must be periodically inspected and maintained. Waste pipes, traps and gulleys must be kept clear of blockage and regularly disinfected.

Consideration must be given to the fitting of spring-loaded or automatic cut-off taps where children are not immediately supervisable (for example, toilet hand basins). All external taps must be lockable.

The main stopcock must be readily accessible for use in emergency and must periodically be tested to prevent difficulty in operation when necessary.

The following should also be observed:

- Drinking fountains, or any other sources of drinking water, must be kept clean by disinfecting daily using a material suitable for drinking vessels.
- Gulleys must be cleared regularly.
- If lead pipes are fitted in any part of the supply from the service main they must be replaced.

- Hot water must never be at a temperature which will scald children, and the hot water supply must be controlled by a thermostat regulator set for a maximum of 30 ℃.

Standing water used for play is discussed in Chapter 5.

4.6 PERIODIC CHECKING

The requirements and guidance set out in this section provide a basis for the safe provision and operation of the adventure playground site and building. However the monitoring of standards and identification of shortfalls will be required in order to ensure that standards are achieved. This will require regular checking of facilities. Sample checklists are provided at Appendices 6 and 7. These provide a model of good practice for periodic checking.

4.7 LEGISLATIVE BACKGROUND

Part Two of this publication contains summaries of the following legislation which has been referred to in this chapter:

- The Health and Safety at Work Act (1974), and its Regulations;
  - Safety Signs Regulations 1980;
  - Electricity at Work Regulations 1989;
  - Control of Substances Hazardous to Health (COSHH) Regulations 1994;
  - Workplace (Health and Safety) Regulations 1992;
  - Manual Handling Operations Regulations 1992;
  - Personal Protective Equipment at Work Regulations 1992;
  - Management of Health and Safety at Work Regulations 1992;
  - Gas Safety (Installation and Use) Regulations 1994;
  - Construction (Design and Management) Regulations 1994;
- Unfair Contract Terms Act 1971;
- Fire Precautions Act 1971, amended by the Fire Safety and Safety of Places of Sport Act 1987;
- Safety Signs Regulations 1980;
- The Occupiers' Liability Acts 1957 and 1984;
- Food and Environmental Protection Act 1989 Part III – The Control of Pesticides Regulations 1986;
- Furniture and Furnishings (Fire) (Safety) Regulations 1988;
- Food Safety Act 1990;
- Data Protection Act 1984.

4.8 FURTHER READING

The following are the main source documents for this chapter:

Dennis, M. (1992) *Tolleys Health and Safety at Work Handbook*, ROSPA.

Department of Health (1989) *The Care of Children, Principles and Practice in Regulations and Guidance*, HMSO.

Food Sense (1990) *The Food Safety Act 1990 and You: A Guide for Caterers and Their Employees, Food Sense.*

Food Sense (1992) *The Food Safety Act 1990 and You: A Guide for the Food Industry, Food Sense.*

HAPA Information Sheet 3 *Setting up an Adventure Playground*, undated.

HAPA Information Sheet 4 *Designing an Adventure Playground*, undated.

HAPA (1994) Health, safety and adventure play, *HAPA Journal*, **13**, pp. 8, 9.

Health and Safety Executive (1970) *The Food Hygiene (General) Regulations 1970*, Information leaflet, HMSO.

Health and Safety Executive (1989) *Guidance Note GS 23 – Electrical safety in schools (Electricity at Work Regulations 1989)*, HMSO.

Health and Safety Executive (1990) *Guidance Note PM 32 – The Safe Use of Portable Electrical Apparatus (Electrical Safety)*, HSE.

Health and Safety Executive (1991) *Guidance Note GS 50 – Electrical Safety at Places of Entertainment*, HSE.

Health and Safety Executive (1992) *Management of Health and Safety at Work – Approved Code of Practice.*

Health and Safety Executive (1992) *Workplace Health Safety and Welfare Approved Code of Practice.*

Home Office (1994) *A Fire Survival Guide: What To Do if Fire Breaks Out in Your Home*, Home Office, Fire Safety in the Home leaflet.

ILAM (1995) *Accessible Play* (HAPA).

ILAM (1995) *Safety in Indoor Adventure Play Areas*, ILAM/NPFA/ROSPA.

Kids' Clubs Network *Food Safety and Kids' Clubs*, undated.

Home Office (1992) *Wake Up! Check Your Smoke Alarm*, Home Office information leaflet.

Home Office (1994) *Protect Your Home From Fire*, Fire Safety in the Home leaflet.

The Children Act Guidance and Regulations, Volume II – Family Support, Day Care and Education Provision for Young Children, HMSO 1991.

Chapter 5

Structures and play features

5.1 INTRODUCTION

A unique feature of adventure playgrounds has been the expectation that children will participate in the development and modification of play structures. The construction provides opportunities for digging, building and creating. The results are often dramatic and add variety and challenge which enhance the play value of the site. It also promotes a sense of belonging and ownership among the children, which has important consequences for the children's view of themselves and the neighbourhoods in which they are growing up.

Where play structures are designed, constructed and maintained using the good practice guidance set out in this book, they will, as far as is reasonably practicable, be safe for those who use them. There are no legal reasons why structures cannot be built and used on adventure playgrounds.

Adventure playground structures offer children challenge, excitement, an environment for imaginative play and an opportunity to develop their strength, co-ordination and agility. They are a genuine challenge against which children can test themselves and where the risks are clearly identified.

Where the skills and resources needed to design, build and maintain structures are not available, some adventure playgrounds have adapted by providing structures for children, either built exclusively by staff or other adults, or bought in from manufacturers of playground equipment. Although these playgrounds continue to involve children by giving them choice and control of a range of other aspects, opportunities to contribute to the building of structures have been lost. This has taken away from children opportunities of interaction with the environment and use of tools in a proper way which is as important to them now as when adventure playgrounds were first introduced.

5.2 PLANNING AND LAYOUT

An adventure playground should aim to provide a variety of opportunities for play. The precise balance between the different elements will depend upon the needs of the users, including those with disabilities and/or special needs. In planning the layout of the playground, space should be provided for a variety of activities – for example, major play structures, building by children, games areas, separate areas for fire, for sand and water play, and an area for younger children. Changes in level, landscaping, planting and areas for keeping animals will enhance the quality of the play environment.

PLAYLINK recommends that an initial risk assessment is undertaken prior to deciding where to locate play structures and other features or when amending the site layout. The risk assessment will provide the context for deciding how the standards and guidelines that follow are to be implemented in each particular case. It is not the case that all playgrounds must comply with uniform safety requirements. However it must be ensured that all reasonable steps have been taken so that children are reasonably safe when using the playground.

Because the adventure playground is an evolving environment it is also important at this stage to consider the potential for the expansion of any of these activities. Where difficulties arise, or where the necessary expertise is not available, advice on the allocation of space and layout should be sought from organizations or expert individuals with experience of these problems. PLAYLINK provides such an advisory service.

The overall layout of the play features must take account of the flow of children into and within the playground, while also catering for a wide range of age-groups and, when possible, children with special needs.

For example the particular needs of younger children should be considered, and it may be necessary to plan for them specifically.

The area around each piece of equipment requires consideration as part of the safety of the equipment. Where it is appropriate to keep it clear to enable children to use the equipment without the risk of collision, an area of 1.8 m around static equipment and 2.4 m around moving equipment is a good starting point when considering what is best for your particular site. A judgement on whether this amount of space is required in each situation will take into account a number of factors concerned with the way the structure is used, movement patterns within the site and evidence of previous accidents. A recorded risk assessment may be the best way to make this assessment.

The positioning of different features within the site can effectively control the flow of users. Structures and areas for specific activity must be sited with reference to site access points and building access. The layout of features must also take into account the possible patterns of supervision.

It is important to observe evolving patterns of movement and of play which develop as they may be quite different from those envisaged during the planning process. Adaptations should be made as necessary.

Play and movement patterns around different structures or activity areas should not intersect – for example, points where children get on and off swings, aerial runways and other moving equipment must not pass over or across the lines of use of, or access to, other items of equipment. Generally, structures or play features should be sited so as to guide flows of activity away from permanent elements of the site, such as the sand area, games area or garden, where activities are more static. Some structures may be sited with the intention of breaking up thoroughfares. Alternatively, natural features of the site, or appropriate landscaping, can serve this function.

Certain structures and activities, such as rope-swings, slides, aerial runways and fire and water areas, are potentially more risky than others and should, therefore, be kept under close supervision. They must be so positioned that a clear view is available from the building and other focal points of the playground.

Some play structures will need to be sited away from overhanging obstructions.

The overall layout of the playground, and the design of individual structures will be changing continually. It is the responsibility of the workers to ensure that the children are made aware of all major changes to be undertaken, and that all necessary measures are taken to avoid accidents resulting from work in progress or completed alterations.

Play areas and structures which are to be used during hours of darkness must be appropriately lit and in some cases floodlighting will be essential. It must be borne in mind that the fewer the number of independent light sources used, the greater will be the unacceptable risk of potentially distracting shadows.

Structures and landscape trees must be well clear of perimeter fencing or the fencing must be designed in such a way as to prevent them becoming a hazard to play activities and to avoid uncontrolled entry to and exit from the site by improvised means.

A plan of the site must be made and kept up to date. This will identify all the elements, dates of installing structures, services, depth of excavations, drain channels and soakaways, old foundations and concrete rafts, infilled/backfilled areas, unused drains, access points, surfacing types, fencing, trees, water table, potential hazards below ground level etc., and will include plans, drawings, maintenance details, and give reasons for positioning which have a bearing on future developments.

Alongside the site plan there should be descriptions of the order and method of construction of each structure, and any additions/alterations to the original structure. This information will assist in the dismantling process.

5.3 SURFACES

Different surfaces are required for serving different functions on adventure playgrounds.

In certain circumstances the use of impact absorbing surfaces can reduce the severity of injuries.

These surfaces have been developed and are manufactured and installed to standards concerned with the reduction of only one type of injury – life-threatening brain injury arising from falls from equipment. They cannot be assumed to limit any other injuries. Research shows that life-threatening brain injuries caused by falls from play equipment are very rare.

PLAYLINK recommends that each structure be considered individually in order to determine whether it is appropriate to lay an impact absorbing surface. In considering both structures with moving parts or static structures, a risk assessment will assist in making decisions appropriate to each particular site.

5.3.1 Bark

Bark and wood chip require daily inspection for contamination, regular raking and periodic topping up in order to maintain the appropriate depth (300 mm is the minimum depth). If it is used under and around moving equipment it will require frequent raking back to refill eroded areas. Not all bark or wood chip is suitable as an Impact Absorbing Surface, and suppliers should be asked to provide a certificate of test in accordance with BS 7188. It should be free of twigs or other sharp objects.

If bark or wood chip has been set alight, once the fire has been extinguished the contaminated material should be removed to a safe place for a period of not less than 24 hours as retained heat may induce spontaneous combustion.

5.3.2 Sand

Sand is suitable both as an impact absorbing surface, and as a play medium. These different purposes however require different sand, round granular sand as an impact-absorbing surface and angular sand for playing with. Impact-absorbing surface sand should be purchased as washed sand and contain no particles more than 3 mm in diameter. It should be laid and maintained to a depth of 300 mm, and be subject to the maintenance routine described above for bark.

Details on sand for play are given in Chapter 5.

5.3.3 Rubber

Rubber impact absorbing surface can be in the form of wet-pour or rubber tiles. They are not usually appropriate for use on adventure playgrounds where change and flexibility are required. Once laid they must be inspected for wear and trip points, and kept free from debris. They are more expensive to install than loose fill materials.

5.3.4 Carpet systems

Carpet impact absorbing surfaces consist of sand retained within a carpet envelope. These should be top dressed with sand according to the manufacturers instructions.

All impact absorbing surfaces should have been tested in accordance with the requirements of BS 7188 and meet the recommendations of BS 5696 Part III in respect of impact absorbing surfaces. They should have the capacity to absorb the impact of a fall from the highest accessible point of the structure concerned. The supplier will advise on this. Generally they should be laid under and in an area extending at least 1.75 m around structures. In considering how far they should extend it will be necessary to consider factors such as the proximity of other structural features etc. They should be laid and maintained according to the manufacturers' instructions, and must be kept free of all hazardous litter or debris.

5.3.5 Hard surfaces

Access routes for both pedestrians and vehicles should be surfaced with an appropriate hard material to aid access and drainage, resist wear and give a safer footing for people moving materials.

5.3.6 General

All surfaces must be maintained in good condition, well drained and kept as free of mud as possible. Mud is a deterrent to some children using the playground and a possible hazard. Particular care must be taken to ensure that access points to structures are kept as free of mud as possible, to minimize the unacceptable risk of accidents caused by slipping on mud deposited on the structure. An area may be provided specifically for mud-play.

5.4 STRUCTURES

As adventure playgrounds developed, playworkers and children began to build large climbing structures, swings and slides. Their purpose was to replace some of the opportunities for adventurous physical play increasingly being denied to children in their local neighbourhoods, and to promote a sense of belonging and an ethos of participation and ownership.

Structures should allow children to take risks and attempt more difficult tasks as their skills increase, without exposing them to hazards. The difference between risk and hazard is important in this context. When taking risks, children undertake a risk assessment of the situation concerned through the application of their knowledge and experience, and as a consequence of that assessment make decisions concerning their own safety. Hazards exist where the application of their knowledge and experience cannot provide the information necessary to make a proper assessment, and they are therefore unable to make realistic decisions concerning their own safety. For example, a girl may take the risk of climbing a ladder to the top platform of a structure. Through the application of her knowledge of ladders and experience of using ladders she has decided she will be safe to do so. However this risk becomes a hazard if, for example, unbeknown to her the ladder has not been fixed to the structure. As she is not able to take account of that information when making her decision she has not been able to make a realistic assessment of her own safety, and she has been effectively endangered by the playground.

Structures must therefore be designed, constructed and maintained in a way which ensures that users will not be exposed to hazards because of unacceptable risks that they cannot foresee. Well-designed, correctly built and maintained self-built structures are no more hazardous than other play facilities when they are under the supervision of skilled playworkers.

5.4.1 Design

The design of individual structures must be considered in the light of anticipated use and reflect the desires, needs and ability of the users. It is obviously not possible or desirable to be prescriptive about the detailed design of individual structures, since it will depend upon such variables as the size and form of the site, the materials available, the relationship with other structures and the skills of the builders. However the following are important general points which must be born in mind.

Strength

It is difficult to specify the precise strength of the materials to be used in adventure playground structures. In general the use of structures will be very intensive and, therefore, those involved in construction must over-specify all components and fastenings, particularly where recycled materials are used. Timber with annual growth rings close together such as Scots Pine and Norway Spruce should be used. Other softwoods suitable for building construction include larch, European redwood, douglas fir and western hemlock. Suitable hardwoods such as ash, European, Japanese and American oak, European beech, aformosia, Brazilian mahogany and teak are preferable where the timber is subject to hard wear. Hardwoods should be from renewable sources.

Stability

Each structure must be integrally stable. The continually developing nature of adventure playground structures may mean that, if structures are dependent upon other structures for support or stability, modification or dismantling may be difficult. The correct installation of uprights and of diagonal bracing will ensure stability in most cases.

Access

Each structure must be provided with sufficient means of access. Stairs, ladders, ramps, nets or other means of access onto the structure or up and down within it must be solidly made and anchored and suitable for their purpose. A variety of different means of access will add to the attraction of the structure as well as avoiding overcrowding. Means of access must be designed so that they are easy to get onto or off and appropriate to the flow of activity on and around the structure: for example, a full-width ramp at the end of a long walk-way may be less hazardous than a ladder. Staircases and ladders should generally have secure fencing on every open side and a substantial handrail on at least one side.

The access needs of children with problems of mobility should be considered. Ramps are vital for wheelchair access. These should be inclined at no more than 1:12 on a ramp of 5 m or less in length, and 1:15 on a longer ramp. The minimum width should be 1.2 m with overtaking places 1.83 m wide. Steps of 300 mm wide rising at an angle of 45 degrees will provide suitable access for the majority of children with mobility problems. Strong handrails should be fitted to ramps and stairs. Other means of access include nets, rope ladders or climbing walls. Walkways between various pieces of equipment can allow continuous movement and enhance play value.

Loading and crowding – excessive stress on a part of a structure caused by overloading can be minimized by careful design, by narrowing walkways or spacing access points. Care must be taken, however, to avoid creating bottlenecks.

5.4.2 Hazards

Playground structures are a focus for active play by large numbers of children and young people. It is important throughout the design and construction process to maintain awareness of potential hazards posed by the structure itself. Particularly where structures are used as, or cross, or are adjacent to, thoroughfares, there must be adequate height below structural members to avoid the possibility of head injuries to children running through, there must be no protruding members or fixings likely to cause injury or to snag clothes as children run past, and no cross members at low levels likely to trip children.

5.4.3 Traps

BS 5696 Part ll provides good advice in respect of traps. It identifies two types; clearances which risk trapping fingers, hands, limbs or head, and wedge traps formed by an acute angle between two or more adjacent parts that converge in a downward direction at or above 1m from the ground. At the time of publication of this document the British Standards Institute was due to publish a Product Assessment Specification PAS 018. This identified other types of trap; whole body, head/neck head first, arm and hand, leg and foot, clothing and hair.

Clearance traps can only be certainly identified through the use of test probes constructed to meet BS 5696 dimensions. These are to be superseded by probes to the dimensions set out in PAS 018. Their use is recommended as an integral part of any pre-use safety inspection of structures.

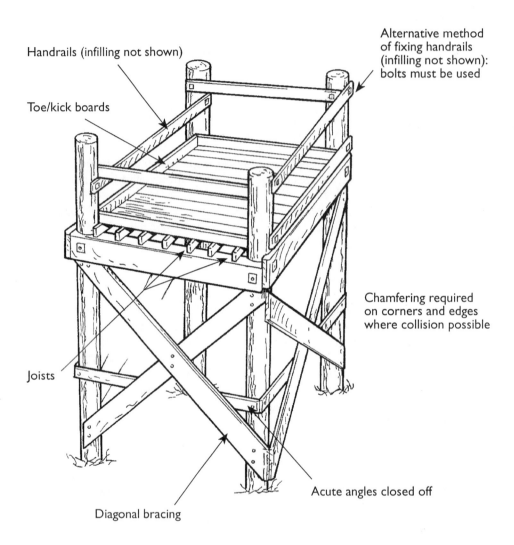

Figure 1. Structural elements (1).

Wedge traps can be identified without specialist equipment. All angles formed by diagonal bracing, or other structural members, which create a wedge trap should be closed off (see Figure 2). PAS 018 will provide further information on wedge traps.

At the time of publication, a draft of the proposed European Standard *prEN 1176–1 Playground Equipment – Part 1: General safety requirements and test methods*, described possible entrapment situations as follows:

whole body, head/neck head first, head/neck feet first, arm and hand, leg and foot, finger, clothing, and hair

and proposes a range of measures to avoid them. Once published the completed standard will provide additional information and guidance in this aspect of structure building.

Design ideas should be talked over with children who use the site, and their ideas and suggestions should be reflected in the overall design. Consultation through pictures, drawings etc. is recommended. The finished plans should aim to provide opportunities for children to become directly involved in the construction in a way appropriate to their abilities.

Design processes will involve preparation of an overall sketch plan of the proposed structure and preparation of detailed scaled drawings of fixing details and other areas as required.

The design of playground structures will, in addition, require an assessment of risk to those who construct and maintain them and those who use them. Information on risk assessment is given in the section below on construction.

5.4.4 Construction

The Management of Health and Safety at Work Regulations 1992 places on employers a requirement to undertake an assessment of risks arising from their undertakings. The design and construction of playground structures will require such an assessment concerning the risks to those who construct and maintain them and to those who use them.

A description of an assessment process is outlined in Chapter 2. In brief it entails an identification of hazards to health and safety which exist, the risks that the hazards pose to employees and others, an identification of precautions already taken, an assessment of the level of risk and, if the risk has been identified as medium or high, identification of additional measures necessary.

In the context of the design and construction of playground structures, the assessment identifies potential hazards to the builders and users, before a final design or construction process is settled on. This will ensure as far as possible that all risks are identified and proper precautions taken in advance of the actual construction and use of the structures.

The risk assessment must include consideration of:

• the safety of those who construct it and those who will use it;

- the siting of the structure and its relationship to other structures and playground features;
- availability and suitability of the materials required, especially whether all components will last for the anticipated lifetime of the structure;
- availability of the tools and equipment required;
- access to appropriate skills required;
- the labour required, including how children can be safely involved in the construction;
- the sequence of operations;
- safety procedures in operation;
- future inspection and maintenance implications.

A further component of this risk assessment can be to seek, as part of the design process, the views of others with experience and knowledge in this area.

In the case of major items, planning the construction must include:

- scaled drawings showing plan, front, back and side elevations, and sections including underground sections as appropriate;
- the preparation of a detailed schedule of materials and tools necessary;
- a detailed sequence of operations, programmed in distinct stages with regard to available labour and the general operation of the playground;
- training/instruction processes required for those involved in the construction.

Any changes to the original plans, designs and/or sequence of operations should be re-evaluated as part of the risk assessment process.

Work should not start until all the materials and equipment necessary for each stage of construction are available and have been checked to ensure that they are suitable and in good condition.

5.5 SAFE PRACTICES IN BUILDING

This section is concerned with good practice techniques for building on adventure playgrounds. It relies on information from the legal framework documents listed at the end of this chapter, and the following Health and Safety Executive Guidance notes:

- *Accidents to Children on Construction Sites*, GS 7 (June 1989)
- *Safe Use of Ladders, Step Ladders and Trestles*, GS 31 (reprinted July 1993)
- *Tower Scaffolds*, GS 42 (reprinted March 1993)
- *Training and Standards of Competence for Users of Chain Saws in Agriculture and Forestry*, GS 48 (December 1990)

5.5.1 Cleanliness, falls and falling objects

The Workplace (Health and Safety) Regulations 1992 include requirements for the prevention of falls and falling objects and cleanliness of a workplace. In the context of construction safety this means:

- all reasonable steps must be taken to prevent falls by a person, or a person being struck by falling objects, by:
  - appropriate use of ladders steps and scaffold towers;
  - provision of protective fencing, where substantial work is taking place over 2 m off the ground;
  - storage of materials so as to prevent them falling;
  - and, where the risk is unavoidable, the provision of appropriate personal protective equipment such as hard hats, safety harnesses etc. and the training required to use them properly.

Waste materials must not be allowed to accumulate on workplace floors or traffic routes by clearing up as work progresses and storing/disposing of waste in a suitable way.

5.5.2 Manual handling

The Manual Handling Operations Regulations 1992 require risk assessment to be undertaken to identify possible risks from manual handling, and an appropriate system of work be put in place to avoid those risks. In the context of construction safety this means:

- manual handling must be avoided wherever possible by:
  - forward planning to avoid or limit the need to lift materials
- the risk of injury should be reduced by:
  - using such alternative means of moving materials as is appropriate, including trolleys, hoists, rollers, lorry loading crane and/or hoist;
  - providing training in lifting and handling techniques;
  - ensuring that sufficient number of suitable workers are available;
  - ensuring that the surroundings are suitable, especially the floor or ground, the lighting conditions and the space available.

5.5.3 Personal protective equipment

The Personal Protective Equipment at Work Regulations 1992 concern the provision of personal protective equipment (PPE) as a last resort in the prevention of risks. In the context of construction safety this means:

- an assessment of what PPE is required for the construction or part of construction and PPE provided, with information and training about its use. This could include:
  - eye protection to guard against impact from drilling, cutting, sawing or nailing, splashes from liquid droplets, dust from cleaning, sawing and sanding;

- hard hats for use in construction areas, particularly for work on or underneath elevated workplaces;
- footwear for protection against falling objects, piercing from sharp objects on the ground and chemical/oil spillage;
- hand and arm protection for manual handling, keeping hands warm and supple in cold weather, when handling hot and cold materials, and when in contact with chemicals;
- safety harnesses for working at height;
- ear protection for noisy processes.

This list is not exhaustive and PPE may be needed for other purposes.

5.5.4 Electrical equipment

The Electricity at Work Regulations 1989 aim to ensure that electrical equipment is constructed, protected and used to prevent risks to health and safety. In the context of construction safety this means:

- electrical equipment must be safe – this is covered in detail in Chapter 6;
- a higher standard of protection is required where electrical equipment is used outdoors, by:
  - the use of cordless equipment wherever possible;
  - the use of residual current devices at the source of supply or close to the source of supply, which are routinely tested by means of the test button before each occasion on which they to be are used;
  - the use of low-voltage equipment, with current correspondingly transformed;
  - the wearing of PPE.

5.5.5 Holes and excavations

Excavations are necessary for the construction of some structures and play features. However there are unacceptable risks arising from the sudden collapse of unsupported or, inadequately supported, sides of excavations in which people are standing/working. (Information on these risks is given in R W King and R Hudson (1985) *Construction Hazard and Safety Handbook*, Butterworth.)

Before an excavation is commenced a ground investigation must be undertaken. This will assess the likelihood of collapse, and identify any other potential hazards. The investigation must consider:

- the consistency of the soil, including moisture content;
- the likelihood of flooding;
- vibration from passing traffic which could loosen the soil;
- the effect the excavation may have on ground stability and adjacent buildings/structures;

- the presence of underground services;
- the likelihood of people entering the hole.

King and Hudson (see above), are adamant that sides of all excavations over 1.2 m deep (save those in hard rock) in which people may stand/work, must be protected in some way. They describe two methods:

1. inside mobile shields or trench supports made from either steel or (good-quality) timber;
2. sloping or battering sides – an angle of 45° is given as the maximum slope for soil in the UK.

They describe the most reliable (and expensive) method of protecting excavations from collapse as being the driving of interlocking steel piling on either side of a trench prior to excavation. Where excavations are hand-dug, the sides must either be battered or supports installed while the excavation is in progress. When a mechanical excavator has been used, a metal frame or cage is required to be inserted in the pre-formed hole.

For adventure playgrounds where these methods of protection may be unachievable, it is recommended that the construction method for which the excavation is required is designed in such a way that it is unnecessary for people to stand/work in the excavation.

For all excavations, soil removed must be retained in a way that prevents it from falling back in to the excavation. If the material is stored as a soil heap near to the excavation, there must be a distance of at least 0.6 m between the edge of the heap and the excavation.

Open excavations must be securely boarded over or fenced when not in use (see fencing and security below). Excavations over 2 m in depth must be filled or securely covered. If leaving them open cannot be avoided, edges must be fenced by a barrier of chestnut paling or similar to a height of at least 1 m.

5.5.6 Fencing and security

Accidents to Children on Construction sites, GS 7 (June 1989) sets out precautions to prevent children from being injured on construction sites. These include fencing, guarding of excavations, immobilization of vehicles and plant, stacking and storage of materials, access to elevated areas and safeguarding electrical and other energy sources. These precautions are relevant to adventure playgrounds in general when they are closed (operators of adventure playgrounds have a duty of care to trespassers under the Occupiers' Liability Acts of 1957 and 1984), and at any time while construction is being undertaken.

The Guidance is relevant to adventure playgrounds in the following ways:

1. *General precautions* – children should be informed about opening and closing times, the hazards of visiting the site when closed or of entering areas where construction is under way. Suitable notices must be provided.
2. *Fencing* – all playgrounds must be fenced, the fencing must not be capable of

being easily climbed or undermined and must be well-maintained, gates must be locked when the site is closed, including when staff are on site for construction purposes and children are excluded. Construction areas, or partly constructed structures, must wherever possible be fenced off and clearly marked as out-of-bounds areas, signs warning of hazards understandable by children must be prominently displayed and all temporary accesses to the structure removed when work finishes for the session

3. *Excavations* – excavations over 2 m in depth must be filled or securely covered. If leaving them open cannot be avoided, edges must be fenced by a barrier of chestnut paling or similar to a height of at least 1 m.

4. *Vehicles and plant* – must be immobilized and, if possible, locked in a separate enclosure.

5. *Stacking of materials* – all heavy materials must be stacked in a way which prevents easy displacement, heaps of sand or loose material must be limited in size to minimize the consequences of their collapse should they be undermined by childrens' digging.

6. *Access to elevated areas* – all ladders giving access to incomplete constructions must be removed and locked away.

7. *Electricity* – electrical supplies must be switched off at isolators in a locked building.

8. *Fires* – fires must be extinguished at the end of the working period.

9. *Storage of hazardous materials and tools* – these must be removed to a secure store.

Other practices at closing-down relating to the outdoor area are described in Chapter 5. A checklist of closing-down tasks is provided at Appendix 5.

5.5.7 Use of ladders and step ladders

The Safe Use of Ladders, Step Ladders and Trestles, GS 31 (reprinted July 1993) gives basic safety information for users of this equipment. Guidance which is relevant to adventure playground constructions includes:

- for ladders:

  - check to see if the job can be done in a safer way without use of ladders or steps;
  - always secure the foot of ladders, and wherever possible secure the top also;
  - the stepping-off rung of ladders must be level with the platform to which they give access;
  - the suitable angle for a ladder is 75° to the horizontal;
  - ladders must only be used for the purpose for which they are designed, check with the manufacturer's instructions for uses other than climbing;
  - extension ladder sections must overlap by a minimum of three rungs;

- for step ladders:
  - side loading must be avoided;

- the top tread is not for standing on;
- stays, which stop the steps from spreading, must be in good order;
- only one person must use a step ladder at a time;

- care and maintenance of ladders and step ladders:

  - they must not be dropped or jarred;
  - they must be individually marked and inspected before and after normal use and regularly by a person competent to do so (a person who has practical and theoretical knowledge and actual experience of ladders);
  - they must be stored on racks designed for the purpose which are easily accessible and if possible locked-off;
  - climbing and gripping surfaces must be free from oil, grease or mud;
  - timber items must be checked for rot, decay or mechanical damage and rungs must be checked for looseness;
  - metal ladders must be checked for twisting, distortion, oxidization, corrosion and excessive wear;
  - other checks are for broken rungs, rivets, tie rods, hinges etc.;
  - ladders, especially wooden ones, must not be painted.

The Guidance also gives advice on trestles, steeplejack ladders, ladder scaffolds and roof ladders, which are not recorded in this publication.

5.5.8 Use of tower scaffolds

Tower scaffolds may be used in the construction of play structures, and GS 42 (reprinted March 1993) provides useful guidance in their use. Tower scaffolds normally have four legs, are not normally more than 3 m in length and heights are variable.

Different types are erected in different ways and have different capabilities and capacities and manufacturers' instructions must be followed. If hired, these instructions must be requested from the hirer. Persons who erect them must be competent to do so, taking into account the complexity of the layout of the tower.

The Guidance is detailed and complex and the following items which are relevant to adventure playgrounds should be considered an introduction only:

1. Check and follow the loading capacity as specified by the manufacturer.
2. Ideally they should be erected on level firm surfaces. In any event the surface must be capable of sustaining the total load. Static towers should have metal base plates and timber sole plates.
3. They must be vertical.
4. All components must be checked for broken welds, cracks etc. before erection and properly tightened, fitted and locked.
5. All towers must be braced in all three dimensions with braces and outriggers, according to the manufacturers' instructions.
6. Aluminium alloy towers differ in stability from steel towers, and the manufacturers' instructions must be followed.

7. Working platforms must be at least 600 mm wide and must be properly constructed and fixed.
8. Guard rails 910 mm to 1150 mm high are required on all working platforms from which a person can fall more than 2 m. Toe boards are also required.
9. Platforms must have a safe means of access, always on the narrowest side of the tower, either by vertical ladders internal to the narrow side, internal stairways or inclined ladders, or ladder sections integral with the frame members. Ladders must be lashed to the tower.
10. Unless access is by an inclined stairway, tools and heavy loads must not be carried, but must be hauled up within the confines of the tower.
11. Ladders must not be used to extend the height of a tower.
12. Where the tower is used for jobs such as drilling, where considerable sideways force may result, great care must be taken to ensure that the tower is not overturned and the manufacturers instruction must be followed.
13. They must be dismantled carefully.

The operators of adventure playgrounds where scaffold towers are used are urged to purchase GS 42 from the Health and Safety Executive in order to obtain full details of the correct practice in their use.

5.5.9 Chain saws and alligator saws

Training and Standards of Competence for Users of Chain Saws in Agriculture and Forestry, GS 48 (December 1990) gives a good insight into the implications of using chain saws on adventure playgrounds. It sets out requirements concerning the selection, training and certification of chain saw operators. The use of these items on adventure playgrounds should not be necessary, but, in the event that they are, they must be used only by operators who have a relevant City and Guilds Certificate of Competence for the specific tasks to be undertaken. 'Alligator' type reciprocating-blade electric saws, which are available in 110 volt, are intrinsically safer.

5.6 CONSTRUCTION DETAILS

5.6.1 Height of structures

Generally it is recommended that the maximum height for any platform or walkway should be 5 m, provided it has appropriate barriers.

Guardrails will normally be indicated on any part of a structure where a vertical fall height of more than 2 m is possible. The construction of one or more intermediate levels to reduce the possible fall heights to less than 2 m could be considered. All steps, ladders and other access to levels higher than 2m above the ground must be within the structures, staircases and ladders must have secure fencing on open sides and a substantial handrail on at least one side. See Figure 2.

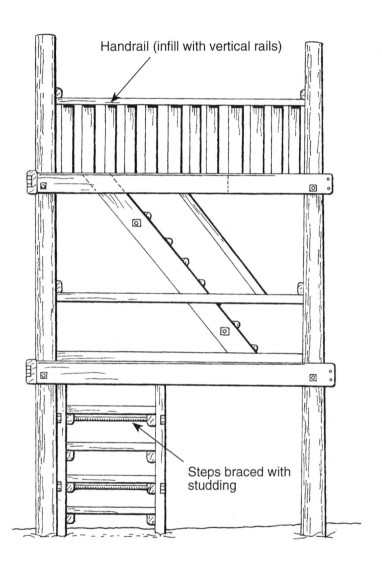

Handrail (infill with vertical rails)

Steps braced with studding

Figure 2. Structural elements (2).

5.6.2 Access

All means of access to structures and to different levels within structures must be solidly constructed and fixed.

It is recommended that any ramped accesses should be at an angle not exceeding 38°. Ramps over 15° should have footholds spaced at distances between 175 mm and 360 mm. Ramps for children with mobility difficulties should be inclined at no more than 1:12 on a ramp of 5 m or less in length, and 1:15 on a longer ramp. There are other special considerations for these ramps (see above).

Generally, stairways should be of a pitch no greater than 45°. The appropriate width of treads and the height between steps depends upon the pitch. Eg stairways with a pitch no greater than 45° should rise 100–200 mm with a going (distance between the front edge of one step and the front edge of the step below or above it) of 220–350 mm. The width of the treads should be 600–1800 mm and their depth should be no less than the chosen going. For steeper stairways up to 55° the rise should be 150–200 with a going of 100–220 mm. The width of the steps should be 280–450 mm, and the depth of the treads should be (if the treads are open) no less than the chosen going, and if closed no less than 150 mm. Steps of 300 mm wide at an angle of 45° will provide suitable access for the majority of children with mobility problems.

The surfaces of steps and ramps must be slip resistant.

Ladders must be of solid construction, with rungs screwed or jointed into the sides. The most suitable angle for ladders is 75° to the vertical.

All access ways must be regularly inspected and any damage promptly and properly repaired. If this is not possible they must be removed, or made unusable, until the necessary work can be completed.

5.6.3 Guardrails

Guardrails will normally be indicated on any part of a structure where a vertical fall height of more than 2 m is possible. A risk assessment will help you decide when, in the particular circumstances of the structure you are considering, a guardrail is required. All guardrails must be a minimum of 1 m high. Infilling below the guardrail must be of solid panels or vertical rails no more than 100 mm apart, which must be securely fixed at top and bottom inside the barrier. Solid panels must extend to floor level. Where this is not possible, or where vertical rails are used, additional solid 'toe' or 'kick' boards must be fitted at floor level to prevent feet or objects slipping beneath the barrier. Horizontal rails must never be used as an inner part of a safety barrier, as children may be tempted to climb on or over them. See Figure 2.

Guardrails must be constructed from materials of sufficient dimensions and strength that they are capable of withstanding the maximum forces which may be exerted on them by children and adults. They must be securely bolted or screwed to main uprights or other structural members, with additional supports to reduce their span where necessary.

The draft of the proposed European Standard Playground Equipment – Part 1: General Safety Requirements and Test Methods was available at the time of publication of this document. This recommends that the cross section of any support designed to be gripped (the hand closed round the entire circumference) must be between 18 mm and 45 mm, and for those which are intended to be grasped (the hand held round three sides) must have a width not exceeding 60 mm.

It is strongly recommended that you consider whether additional barriers be set below platforms at first-floor level which have an underside of less than 1.7 m above the ground. This is to avoid the possibility of children hitting their heads on them when they play round and under them.

5.6.4 Platforms and flooring

Planking or other surfacing must be flush with the supporting joists, even and free of obstacles, intrusions or projections. It must not overhang the outside joists. All surfacing must be firmly nailed down and if damaged, repaired immediately. Until such remedial work is effected, the structure must be made inaccessible.

Leaving gaps of less than 1 cm between planking will allow water to drain away.

Care must be taken in the selection of materials for flooring or other flat surfaces such as access ramps. They must provide a good grip and, where possible, not become dangerously slippery when wet.

5.6.5 Raising main uprights and using structural timbers

Heavy construction work, such as the erection of uprights or major structural members, must be undertaken when the site is clear of children. When these operations are scheduled, a realistic time must be allowed to ensure completion before the site is opened to children.

Telegraph or electricity poles or other heavy timber which are main uprights or structural components, must be erected so that at least 25% of their total length is below ground level, irrespective of the length of the pole. This proportion must be increased if the upright will be subject to severe lateral stress. Uprights and all ground-contact timber must have been treated with an effective preservative before coming onto the site. When the use of secondhand telegraph poles is being considered for main uprights or as structural components, checks should be made to ensure that they are sound, checking for rot, cracks, and other defects.

The use of appropriate mechanical plant for the erection of poles is recommended, and in any event the Manual Handling Operations Regulations 1992 recommend that manual handling be avoided as far as is reasonably practicable, and that if manual handling operations cannot be avoided steps must be taken to reduce the risk of injury so far as is reasonably practicable.

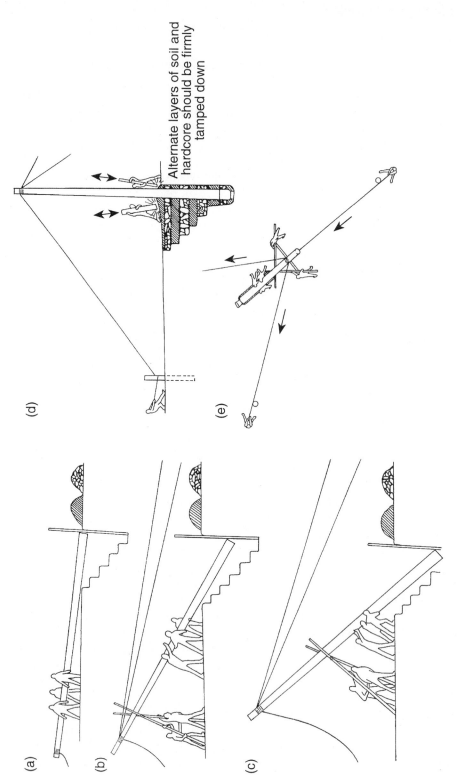

Alternate layers of soil and hardcore should be firmly tamped down

(a)

(b)

(c)

(d)

(e)

Figure 3. Raising an upright.

If the avoidance of manual handling is not possible, for the actual raising of the upright it is necessary to have an adequate number of people to raise them safely.

One person must take strict control of the whole operation. The supervisor must be in a position where s/he can see the whole operation, and must issue all instructions and directions while taking no active part. All members of the team must be thoroughly briefed on the whole sequence of operations and on their own part. If any mechanical plant used is operated by someone not familiar with the procedure, they must be fully briefed by the supervisor.

There are a variety of methods of raising main uprights and individual teams of workers will evolve variations which best suit their circumstances. The method which follows must therefore, be seen as only one possibility. The sequence is also outlined in Figure 3.

Holes for main uprights must be prepared as narrow as possible. Because of the depth needed, if the digging is done manually it will generally be necessary to dig a 'stepped' hole. The bottom of the hole must be rammed firm before the upright is placed. If a hole is more than 1 m deep it must be shored , or widely banked appropriate to the type of soil etc. if persons are to descend into it.

The upright must be placed over the hole and the pole lifted until its base is at the bottom of the hole, and it rests at the angle of the 'stepping' if there is any. If a backboard has been used it must be removed at this point. For a manual operation, at this point in the process the upright must be supported by a suitable 'A'-frame, strong enough to support the whole weight of the upright. Three ropes must next be fixed close to the top of the pole prior to any lifting.

In a manual operation from this rest position the pole may be pushed into an upright position, supported by equal steady pull on two of the three ropes. The third rope is to steady the 'back' of the pole. When the pole is in an upright position, the supervisor must adjust the pole to a true vertical by directing adjustments to the tension on the supporting ropes. (At this stage it may be advisable to increase the numbers holding the ropes.) Once a true vertical is achieved (which is often easier to do by sighting against buildings or other known vertical than by the use of a spirit level or similar instrument, particularly when a pole is tapered), the pole may be held upright. This is done by securely tying off the ropes at the correct distance or by attaching temporary supports (e.g,. timber braces at 45°) while the hole is filled.

Under no circumstances must existing structures be used to provide support, or as anchorage points for ropes, when heavy uprights are being erected.

To secure the upright, the hole must be backfilled with alternate layers of earth and hardcore. Each layer must be no deeper than 225 mm and each layer must be firmly tamped down before the next is added, in order to eliminate pockets of air and water traps (See Figure 3). Concrete must not be used except in exceptional circumstances (such as where the soil is very light). In such circumstances, playworkers should seriously consider the suitability of high structures for the soil conditions. If concrete is to be used, care must be taken to avoid the creation of a water trap at the base of the pole. The lower end of the upright must not be enclosed in concrete, and where concrete extends to the surface, it must be

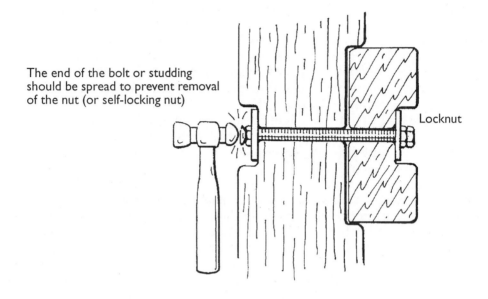

The end of the bolt or studding should be spread to prevent removal of the nut (or self-locking nut)

Locknut

Figure 4. Peining.

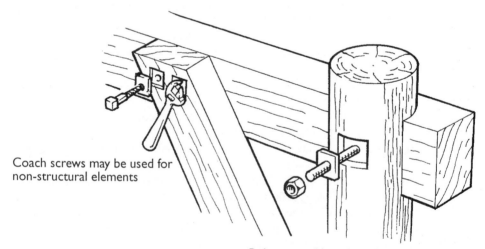

Coach screws may be used for non-structural elements

Bolts or studding should be secured with a nut and a plate countersunk into the wood

Figure 5. Fixings.

contoured to shed water away from the wood. No one should enter an unprotected excavation.

All structural (load-bearing) members must be of adequate dimensions for the possible loading and the width of unsupported spans. All structural joints must be between two flat surfaces. With round sections, notching will be required. This will reduce the diameter and therefore the strength of the timber, so allowance must be made for this at the design stage and when specifying and selecting materials.

All structural joints must be fixed with bolts or studding of appropriate gauge, 16 mm with 50 mm round or square washers are usual, minimum size – 12 mm. All bolted and studded joints must include large washers or (preferably) steel plates, and must be secured with lock-nuts (see Figures 4 and 5) to prevent nuts working loose or being removed. All projecting ends should be cut flush or covered.

All structures must be adequately stabilized by the use of diagonal bracing. Such bracing plays an integral part in the strength and stability of the structure, particularly where the structure is subject to lateral stress and must be of appropriate material of suitable size to resist foreseeable strain. Wedge traps, where bracing meets uprights, must be closed off. See section on traps (above).

5.6.6 Slides and ramps

Along with aerial runways these are among the most popular, and most potentially hazardous features on adventure playgrounds and, therefore, need careful construction, constant maintenance and maximum possible supervision. They can prove extremely difficult to construct to required standards of safety. They must be carefully sited, with attention being paid to providing adequate run-offs, well away from flows of children.

With these difficulties in mind, it is recommended that purpose-built continuous piece, stainless steel slides should be considered as the first choice for slides on adventure playgrounds. These should be installed in accordance with the manufacturer's instructions, should not face south (because the metal surface will get extremely hot during the course of a sunny day) and be inspected daily.

Slides should preferably be sited on embankments.

If it is not possible to use a continuous metal runway, great care must be taken in the selection of construction material. Certain timber will be unsuitable because of a tendency to splinter. Ideally, exterior or marine quality plywood should be selected. Children may be encouraged to sit on or in a suitable item when sliding. The sides of the bed must be built up with close-boarded guard rails set at an appropriate angle and overlapping downwards. Buffers made from a suitable material, such as high-density industrial rubber, may be used on the sides. Run-out sections must be safely sited with regard to other structures and activities. They must be horizontal or slope downwards at an angle of up to $-2.5°$, be as low as practicable and placed on a hard wearing surface with no holes or traps. Run-out sections for slides up to 2.5 m high require minimum of 20% of total slide length, 2.5–5.0 m high require 25% and 5.0 m high and over at least 33.3%.

Entrance areas must provide protection from falling, space for children to settle themselves onto the slide and take off when they choose, and suitable grab rails.

Impact absorbent surfaces are generally required if there is a fall height of 1 m or more. Loose fill surfaces such as bark, are not successful as an impact absorbing surface in run-out areas.

5.6.7 Aerial runways

Aerial runways are exhilarating features of adventure playgrounds. Despite an appearance of danger, they are regularly used with great enjoyment and without incident on many playgrounds. Attention to these guidelines for installation and maintenance will ensure that children can continue to have access them and the challenge they provide. Useful guidance on the design and construction of these items is also contained within the DIN Standard 7926.

Their location must be carefully selected in order to ensure that the runway avoids other activity flows. This usually means placing them near an edge of the site and parallel to it.

The surface below the runway should be hard wearing, as feet will be dragged and the surface heavily worn. Impact absorbing surfaces are generally not necessary where there is a fall height of less than 1 m. The DIN Standard 7926, Part 4 measures the fall height from the surface of the seat on a 'user-seated' ariel runway. 'User-suspended' aerial runways (from handgrips on the pulley block) are not recommended (see below).

Choosing the correct surface is problematic and careful thought must be given to the balance between the need for a hard-wearing surface and the likelihood of injury as a consequence of falling onto it. Loose-fill impact absorbing surface is only suitable if it can be raked back into position as often as it is shifted.

Embankments or mounds at take off and landing points can be used to reduce the fall height of the runway. If the equipment is bought-in rather than self-built, the recommendations of the manufacturers concerning location, installation and maintenance must be followed – not all can be put successfully on embankments. If the equipment is self-built, then particular attention must be given to the requirements concerning angle of descent, slack in the running line and stopping, which are described below.

Aerial runways are major structures and, hence, fairly permanent. Purpose-built towers and platforms must form the basis of the runway – trees must never be used as anchorage points. Towers or platforms must be adequately braced against stress and the running line must be truly aligned between anchorage points.

The take-off platform should be constructed according to the requirements of the individual runway, taking account of the point of take-off, how the pulley will be caught on its return and how the children will travel on the runway. The platform must be neither so large as to get overcrowded, nor so small as to make take-off an unacceptable risk. It should be constructed to allow playworkers to help children with disabilities onto the seat or into a harness, without the unacceptable risk of the child's weight taking them down the runway before they are ready. Generally, if the take-off platform is higher than the recommended fall-height of 2 m, an intermediate platform must be provided for access and suitable guard rails fitted at the take-off level. Safety nets are not recommended, since they are easily vandalized and children may be seriously injured when falling awkwardly.

The use of a rubber, or other impact absorbing material, seat is recommended as safer than using suspension handles. This should be suspended below the pulley block on a rope handgrip. The rope must be securely attached to a closed ring on the block – the use of an open hook closed by 'mousing' (closing with wire or string) is not acceptable. The weight of the user must be directly below the pulley block. Seats must be constructed so that the user cannot come within 1 m of the cable at any time while travelling on it.

A stop must be provided on the take-off end of the cable, to leave the seat within easy reach of the next user on the platform, while preventing the pulley block hitting someone on its return. The cable and pulley block must not be within reach of children using the runway.

The angle of descent must not be so great as to create excessive speed. There must be sufficient slack in the running line to enable users to touch down safely on the ground by foot. (See Figure 6). If necessary, an angled ramp can be provided.

Sudden stops must be prevented. This is best achieved through adjustments to the slack in the cable and the height of the end support so that the pulley slows and comes to a gradual stop. In any event the pulley must stop a sufficient distance from any construction which if struck is likely to cause injury to the user, and to prevent the pulley from hitting the structure.

Rope should not be used as a running line. Steel cable of 10 mm diameter is common. 13 mm cable is also used.

Steel cable may be seen as a permanent installation, but will still demand a comprehensive schedule of inspection and maintenance. The cable should be regularly greased to ensure smooth travel of the pulley and to maintain its condition. It must be inspected daily as closely as possible for signs of wear and loose strands ('nibs') and dismantled for thorough examination at least once a year. Heavy protective gloves must always be worn when handling steel cable.

The whole anchorage system must be clearly visible at both ends, with no part of the cable buried. The cable must be safely secured at each end by bulldog grips of a size corresponding to the diameter of the cable used, and which meet the standard of DIN 1142. These must be fitted with the U clip on the untensioned/unloaded end of the cable (see Figure 14).

The number of grips to be used relates to the diameter of the cable. For 10 mm and 13 mm cable four grips must be used for each fixing point. The spacing between the grips should be at least six times the diameter of the cable.

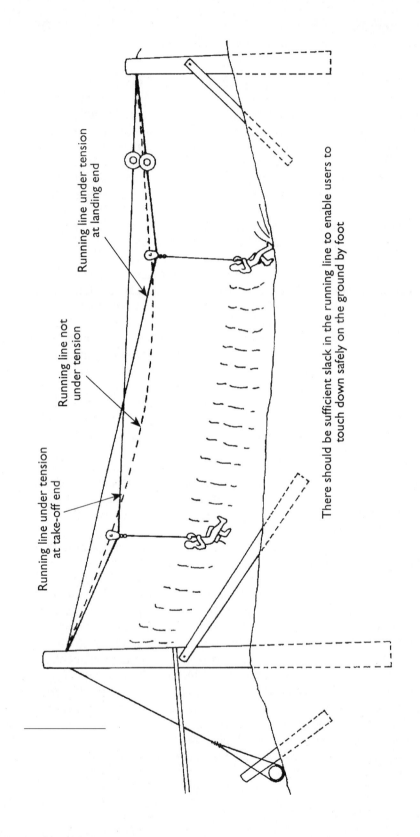

Running line under tension at landing end

Running line not under tension

Running line under tension at take-off end

There should be sufficient slack in the running line to enable users to touch down safely on the ground by foot

Figure 6. Aerial runway tension.

Figure 7. Cable (running line) protection.

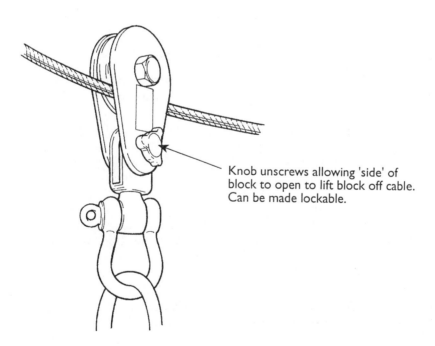

Knob unscrews allowing 'side' of
block to open to lift block off cable.
Can be made lockable.

Figure 8. Snatchblock pulley.

Each nut should be tightened evenly and in turn to achieve the recommended torque setting. For grips used with 10 mm cable, 9 N.m. for those used with 13 mm cable 33 N.m. 24 hours must elapse before the runway is used. At the end of the 24-hour period the torque settings must be checked and the nuts retightened if required. After the first day of use these settings should be checked and if necessary reset. The torque settings should be rechecked when servicing the cable.

In addition, a large turnbuckle, or bottlescrew, may be provided to allow for adjustment of slack, and to take up stretching in the cable. Children must be discouraged from playing on or around the anchorage points. The whole anchorage system, including the bulldog grips, must be inspected daily, with periodic checks of the torque settings.

Wherever the cable is in contact with the structure, the point of contact must be protected by sleeving the cable or by padding the structure with an appropriate material. (See Figure 7).

All pulley blocks must be free-running, in good condition and of the correct size for the cable. The runway must be immobilized when the site is closed. The most effective way of doing this is to use removable pulleys such as snatch blocks (see Figure 8). Where this is not possible, the pulley must be locked off at one end of the runway. It is desirable that removable pulleys should have a mechanism to lock them onto the cable when in use. The purchase and use of pulley blocks specifically designed for play structures is recommended, if they are compatible with the cable used. Some types have a weight-break which stops toddlers using them and prevents their being sent along the wire without a rider. However these are expensive.

An extension to the handgrip to enable easier return of the pulley must be of lighter rope or heavy string. Care must be taken to avoid the handgrip or return string snagging on the take-off tower.

Adventure playgrounds should try to provide a range of seating types which can be easily changed to enable children of all abilities to use the runway.

Procedures for using the runway, based on the safe working load of the pulley, cable and structure, and any design features of the runway which restrict or influence its use, must be communicated to the users, and rigorously enforced. Children new to the adventure playground should be instructed in its use, and a high level of supervision maintained. Where supervision cannot be maintained, and whenever the playground is closed, the runway must be immobilized by the removal of the pulley block, or similar.

Playworkers must test and satisfy themselves as to the safety of the runway and its individual components daily, with a weekly detailed examination of such vulnerable parts as the pulley spindle, and handgrip fastening. Annual inspection by an independent qualified person is recommended.

The DIN Standard 7926 provides detailed guidance on the construction of cable runways which will serve as important information to those proposing to self-build one.

5.6.8 Swings

Ropes and swings must be carefully sited, or provided with barriers to avoid accidents caused by children colliding with these items when in use.

In considering whether impact absorbing surfacing should be installed below swings it should be noted that some types of this surfacing are not suitable for areas where there is constant movement. A risk assessment will assist you to reach a decision.

The area of surface to be covered is calculated as follows.

For each swing seat which is intended to move only backwards and forwards,

Width:

- the width of the surface should extend at least 875 mm either side of the centre of each swing seat, but not beyond the inner supporting legs of the swing.

Length:

- measure the length of the swing chain/swing rope from the point where it pivots at the top, to the surface of the swing seat;
- multiply that length by 0.866;
- and add 1.75 m;
- the surface should extend for that length both in front of and behind the seat, measuring from the centre of the seat.

For each swing seat which is intended to move in multiple directions e.g. cantilever swings,

- use the LENGTH calculation above;
- use the result of that calculation as the radius of a circle with the centre of the seat as the centre of the circle;
- the circle is the area which must be surfaced.

Swing uprights subject to any lateral stress must be of sufficient dimension and strength to take account of all actual or potential stress and must have at least 25% of their overall length below ground level. Uprights must be adequately ground braced, and crossbeams adequately cross braced, to resist lateral stress and movement.

For swings which provide for users to take off from a platform, the positioning of the platform leading edges and take-off points to the arc of the swing are critical. These can be difficult to correctly calculate at the design stage. It is therefore recommended that platforms and take-off points are constructed after the main swing elements, and that a rope or cord suspended from the swing anchorage point is used to check the positioning of platforms and take-off points so that the correct balance between adequate clearance and ease of getting on/off swings is obtained.

Swings must be designed so as to facilitate access to fixings and other parts for purposes of inspection and maintenance, but casual access by children must be avoided. Ropes must be fixed so as to enable swings to be removed or disabled when the playground is closed (see Figure 9).

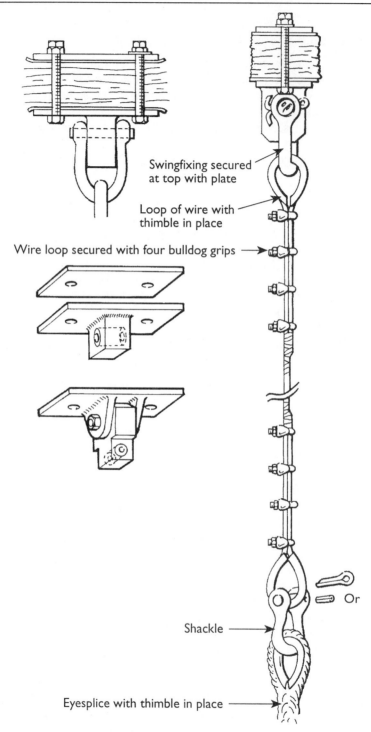

Figure 9. Cable and rope fixings.

Where seats are fitted to swings they must be of an impact absorbent material and free of sharp edges. A mixture of toddler and junior seats in one bay is not recommended.

Swings and other moving items inevitably place increased stress on the structures which support them. Any such structure must be constructed of appropriate materials and suitably braced and supported to withstand these stresses. All fixings for the swing or rope must be of hardened steel and adequate to support the weight.

Swings and ropes must be removed or disabled whenever the playground is closed. When they are replaced, the rope or swing should be replaced in the same position. Colour coding or other marking of individual items may be useful to ensure this. The process of removal must not expose the member of staff to unacceptable risk.

5.6.9 Purchasing purpose-made components

Purpose-made fittings such as aerial runway pulleys, are available for purchase from manufacturers of playground equipment. These components are useful, as they are designed and constructed for similar purposes for which they are required on adventure playground structures. The manufacturers' advice must be sought on compatibility of uses. Purchasers must ask for written confirmation that items have been tested in accordance with the methods described in BS 5696 Part 1 to the standards set out in BS 5696 Part II, or the relevant section of the EU Standard, when published.

5.6.10 Building by children

Structures

Children should be permitted and encouraged to participate in the construction of play structures. Decisions on how and when this can take place will depend upon particular circumstances and the age and ability of the children concerned. The design of playground structures requires an assessment of risk to those who construct and maintain them and those who use them, see above. As part of that risk assessment, consideration must be given to opportunities for the participation of children in their construction and the health and safety requirements that will be required as a consequence. The information contained in this chapter will be of use in this assessment.

Dens and camps

Dens and camps built by children provide very high levels of play value, because of the opportunities for fantasy play, manipulation of the play environment, socialization and co-operative play. These small protected spaces enhance feelings of security and independence. Playworkers should be aware of potential hazards so that they are able to advise children about siting, construction and furnishing.

More elaborate wooden structures intended to have some permanence should be single storey and below 2 m in height. Timber should be thoroughly inspected before being released to children for building. Children should be encouraged to think about floors, roofs and main supports and shown how to make them properly load-bearing. They need to understand the possible consequences of collapse, getting shut in or hurting themselves or others through the use of naked flames or flammable materials for furnishing, and therefore why there are safety rules.

One way of dealing with the potential hazards while continuing to encourage children's creativity might be to help them draw up their own handbook dealing with siting, construction, ventilation, fire risk and so on.

More informal dens may be constructed using large pieces of cloth or arrangements of large moveable pieces of equipment. Planting schemes can also offer a similar experience of small protected spaces giving privacy and independence.

Children should be encouraged to:

- provide adequate ventilation in their camps and dens; the use of plastic or polythene sheeting for waterproofing can restrict ventilation as well as creating a real fire hazard;
- avoid the use of unsuitable materials, particularly for furnishing; these include mattresses and upholstered chairs; these will be a fire hazard and may be, or become, verminous or mouldy.

Children may wish to build tree-houses. The attitude of the workers will depend on the characteristics of the particular site and the trees available. All the principles of safe construction set out in this book must be taken into account. Special care must be taken to ensure structural safety and adequate access. Expert opinion on the condition of the trees should be sought. Particular care must be taken to avoid damage to trees.

Building by children should be checked at least daily and more frequently where there is a lot of building activity. If possible, children's buildings should be sited away from structures built by playworkers.

Holes and excavations

It is useful to have an area set aside for excavations by children.

Before the area is chosen a ground investigation must be undertaken. This will assess the likelihood of collapse and identify any other potential hazards. The investigation must consider:

- the consistency of the soil, including moisture content;
- the likelihood of flooding;

- vibration from passing traffic, which could loosen the soil;
- the effect the excavation may have on ground stability and adjacent buildings/structures;
- the presence of underground services.

Children should not be permitted to dig where a likelihood of collapse has been identified, nor deeper than their waist height.

Digging must take place only in soil which has good natural drainage. Should water collect and not seep away, holes must be filled in. The location of each hole must be easily visible. Any hole no longer in use must be filled in.

The construction and use of all excavations needs close supervision. Danger will exist because of the possibility of sides collapsing, and the tendency of holes to collect water and to provide a receptacle for litter. They must be checked and cleaned daily.

Soil removed from a hole must be retained in a way that prevents it from falling back into the excavation.

Underground tunnels are potentially very dangerous and must not be allowed.

Use of tools

Children must be shown how to use tools safely, but not be allowed to use tools which are too big for them, or too complex, or potentially hazardous to handle safely.

5.6.11 Inspection and maintenance

Inspection and maintenance must be seen as an essential aspect of running an adventure playground, to be carried out continually and systematically, and be properly recorded.

Inspections must be undertaken on a regular basis and carried out in accordance with a written check-list which specifies:

- date and time inspection was concluded;
- items checked and noted as satisfactory;
- the frequency of inspection;
- faults reported and action to be taken;
- the signature of person/s undertaking the inspection;
- consequent action together with appropriate dates.

Suggested check-lists are given in Appendices 6 and 7.

Inspections must take place:

- daily – a recorded inspection – before the playground opens for children, and after it closes, including monitoring cleaning specifications and focusing on items where damage could result in immediate hazards, e.g. cables, steps, etc.;
- weekly – a recorded inspection – inspection of structures and the outdoor area;

- quarterly recorded inspection – playground inspection focusing on the outdoor area, conducted by a competent person from outside the staff team but (if possible) within the organization;
- annual recorded inspection – conducted by an independent expert with an understanding of play.

Inspections must pay particular attention to play structures. Children will use them in many unforeseen and unpredictable ways and inspection must be geared to uncovering all signs of weakness. If a consistent pattern of use appears which is obviously different from that originally intended, then the structure concerned must be adapted to suit that use, or removed.

Children should be encouraged to contribute to the construction and maintenance of the structures and this will be planned for (see risk assessment of structures above). Although provision may be made for opportunities for children to add, e.g. dens to structural components improvised and unofficial additions to the play structures should not be permitted. Regular inspections and maintenance should include the identification of such alterations, and any remedial action required.

Maintenance must take place whenever faults or hazards are identified, or steps are taken to avoid them. The recording of 'near misses' can be a valuable part of this process.

A schedule of preventative maintenance may be drawn up, to be undertaken during periods of less intensive use of the playground.

It may be necessary for playgrounds to be closed at certain times for comprehensive maintenance to take place. Closure must be agreed with the management and all efforts must be made to inform the children and the local community.

The recording of inspections and maintenance is an essential part in the health and safety system. A complete record must be kept of all outdoor structure inspections and maintenance and any development work. This provides a history which informs risk assessment and maintenance scheduling.

Although responsibility for health and safety lies with everyone, the nomination of one person to focus on issues of health and safety and to follow up remedial action required following inspections is recommended.

Time must be programmed into the working day for these essential safety functions.

5.6.12 Dismantling

Both the philosophy of adventure playgrounds and the need for care with regard to safety require that structures are subject to change. The frequency of change will depend on the expressed and observed needs of the children, the safety of structures, the materials of which they are constructed and the resources available. Dismantling of structures is, therefore, an integral aspect of playground work. It can be hazardous and, therefore, a precise, considered approach is necessary. For any major structure, it is generally recommended that the work is carried out in the exact reverse of the sequence of operations used to build it. Records of the sequences of operation should be consulted (see above).

Dismantling major structures should take place when the playground is closed to children. Poles to be removed and timber supports to be lowered must be adequately supported by mechanical means, or if that is not possible, by the use of A frames and ropes (see above). Linking structures should be braced to compensate for reduction in strength and stability. New guardrails must be provided where necessary. As with any other structural work on the playground, the correct tools must be used. Appropriate protective clothing and equipment must be used.

All excavations not to be re-used must be back-filled and firmly rammed immediately. All debris resulting from dismantling must be disposed of and the ground raked clear. If a hole is to be re-used, it must be safely and securely covered.

5.6.13 Landforming

Variations in height and level enhance the play value of an outdoor area. Where variation of contour is to be achieved by landforming, careful consideration must be given to the effect of the work upon the rest of the site. Plans should take into effect the natural drainage of the site and the effect of new contours on that drainage. It is also important to be aware of the effect on the supervision of the site if landforming creates visual barriers. Detailed planning is necessary and expert advice should be sought. Steep slopes (beyond the angle of repose of the soil) should be avoided.

While the work is in progress it may disrupt activities. Access will be required for heavy plant, and children will have to be kept away from the areas of work. It may be necessary to remove parts of the play structures to provide access. When the work is completed it may be necessary to isolate the new areas for a period to allow settling and growth of new grass. Surface reinforcement materials will enhance the strength of new surfaces.

New features must be completely covered with a minimum depth of 150 mm of topsoil. Hardcore or other ballast material must not be exposed.

5.6.14 Sand

Sand is an important resource for children's play. It provides opportunities for manipulative, tactile, fantasy and physical play to take place. Every adventure playground should provide for sand play, and not just for younger children.

Sand play areas must be clearly defined, separate areas, sited away from structures, their run-out points and movement patterns. As sand is to a certain extent self-cleaning when exposed to sun, rain and wind, sand play areas should be situated in a sunny and open part of the playground. A distinct, raised perimeter is recommended since it will not only clearly define the boundary, but may also help to keep out litter, leaves etc, and to retain the sand.

A low wall surrounding the sand area is ideal. A height above the ground of 450 mm will permit its use as a seat for children using the sand. Provision for access by wheelchairs can be provided by a sloped entrance area. The sand should be approximately 300 mm below the top of the surround.

The sand area must be soundly constructed on freely-draining foundations and provided with a soak-away or an outlet to a surface-water drainage system.

Sand should be angular and therefore binding when damp, and non-staining. Silver sand, or sea sand, though expensive, are recommended. Where these are not available, thoroughly washed builder's sand is adequate. The recommended particle size is 1.5 mm or less. Sand should be between 380 mm and 450 mm in depth.

Sand must be thoroughly raked at least daily to remove rubbish (special care must be taken to ensure that no broken glass remains) and disinfected at least once a week with a dilute solution of disinfectant used in accordance with manufacturers' instructions. Periodically it must be turned over to its full depth to avoid fungal or bacterial growth. The level should be topped up as required, and the whole contents replaced each year.

Sand areas should be covered when not in use with a wire mesh or similar open cover rather than a solid one. It should be easily removed and replaced, and be securely stored when not in place.

5.6.15 Fires

Fires have a fascination for children, but the disappearance of the open fire in the home has lessened the opportunity for direct experience of fire. Playground fires provide not only that experience but also serve as a focal point for social contact. With proper supervision children learn to understand the unacceptable risk of fire – and thereby learn to respect it.

If fires are provided on adventure playgrounds there must be a properly established, recognized area for fire, with a hard, positively non-combustible base, which can be dampened down in periods of dry weather. Fires must be built at ground level. Fires dug into the ground are unacceptable. It is particularly important to ensure that there are no underground services near the fire base. Fires may be built in metal containers, provided they are effectively stabilized. Avoid concrete pipes since these can explode. It is extremely important that children are kept at a safe distance (see below) since the temperature of the surface of the container may not be immediately apparent. Fire areas can fulfil the dual roles of providing for occasional fires and also for barbecues and cooking.

Fire areas must be sited away from areas of high activity and from structures, buildings, fencing and vegetation. Consideration must be given to those living nearby. Improvised seating of non-combustible material, such as earth banking, around fires, and at a minimum distance of 2 m from them reduces the unacceptable risk of children accidentally or intentionally running past or leaping over them. Railway sleepers, which are often heavily impregnated with creosote, must not be used for this purpose.

Smoke is dirty and can often be a nuisance. The burning of certain materials, such as foam upholstery and treated timber, will generate smoke which is highly toxic, and is illegal. The direction and strength of the wind must be considered before lighting the fire. Playgrounds should find out if they fall within a clean-air zone and should contact the local authority concerned in order to obtain information relating to possible bye-laws limiting fire.

It is preferable to start fires with paper and dry kindling. In certain circumstances fire-lighters may be used. Only small quantities of fire-lighters must be stored on site, and they must be kept under lock and key. On no account must petrol, paraffin or other highly flammable substances be used either to start a fire or to keep it burning. When a fire is burning it must be properly supervised by a responsible adult at all times who has the appropriate personal protective equipment.

Rubbish must not be thrown on fires. Certain materials may explode, or give off toxic fumes, or create molten rivers of flame. Children must not be allowed to add any items to the fire, unless approved by the adult supervising the area. The following items can be particularly dangerous in and around fires:

> asbestos, aerosol containers, concrete slabs or slate, bottles and other glass items, pipes, batteries, tin cans, foam rubber, tyres, plastics, plastic material, chemicals, gas bottles, empty containers of flammable agents or weedkiller, or unidentified substances.

A supply of water, sand or loose soil adequate to extinguish the fire must be kept near-by, and a fire blanket, especially if it is a cooking fire. If such means are not available, then no fire may be built. Clean, cold water is also essential for immediate first-aid in case of accidents.

Any fire must be completely extinguished when no longer needed and not left to burn itself out. A fire must not be left unattended until it is completely out. The fire must be repeatedly doused with water and raked until it has been completely extinguished. Special care must be taken to ensure that containers and/or firebases are completely cooled.

5.6.16 Water

Water play is important to children. Water provides opportunities for physical and mental stimulation, particularly for children with disabilities.

Outside water-play pools are a feature of many adventure playgrounds and the following health and safety points relate to them:

- the pool should be situated in a safe and prominent position, easily visible from most points of the playground;
- no pool must exceed 600 mm in depth;

- once filled with water, it is essential that it is properly circulated (pumped) and filtered so that it remains free of bacteria and algae; these must be regularly tested for;
- it must always drain away freely into the mains system;
- it must be cleaned at least once a week; stiff brushing will be necessary;
- pumps, filters and drainage pipes must be regularly cleaned to prevent the build-up of sludge etc.;
- pool edges and sides must slope gradually into the centre, thereby eliminating falls into the water and enabling easy access;
- bases must be strong and resilient; concrete bases with a smooth covering of sealant are often used;
- no water must be left in an unsupervised pool;
- constant checking of the water is necessary for glass, nails, stones and other dangerous objects; these must be removed immediately.

Inflatable pools should be sited appropriately and emptied daily. They should be of sufficiently sound construction for the type of use and age-groups of children using them.

Any areas of water on the playground must be constructed and maintained to enable them to be drained while the playground is closed.

Water-play features which combine sand and water-play are fun and stimulating. Care must be taken to ensure that the drainage provided can accommodate the sand washed through. The design of the playground can provide places for puddles to form, to encourage informal water-play.

5.7 LEGISLATIVE BACKGROUND

Part Two of this publication contains summaries of the following legislation which has been referred to in this chapter:

- The Health and Safety at Work Act (1974) and its Regulations;

  - Electricity at Work Regulations 1989;
  - Control of Substances Hazardous to Health (COSHH) Regulations 1994;
  - Workplace (Health and Safety) Regulations 1992;
  - Manual Handling Operations Regulations 1992;
  - Personal Protective Equipment at Work Regulations 1992;
  - Management of Health and Safety at Work Regulations 1992;

- The Occupiers' Liability Acts 1957 and 1984;
- The Consumer Protection Act 1987.

5.8 FURTHER READING

The following are the main source documents for this chapter:
Day, C.L. (1991) *Knots and Splices*, Adlard-Coles, 1991.

Draft of the proposed European Standard prEN 1176–1 *Playground equipment – Part 1: General Safety Requirements and Test Methods.*

HAPA Information Sheet 4, *Designing an Adventure Playground.*

HAPA Information Sheet 5, *Play Structures and Play Equipment.*

Health and Safety Executive (1989) *Guidance Note GS 7 – Accidents to Children on Construction Sites*, HSE.

Health and Safety Executive (1989) *Guidance Note GS 23 – Electrical Safety in Schools (Electricity at Work Regulations 1989)*, HMSO.

Health and Safety Commission (1990) *COSHH: guidance for universities, polytechnic and colleges of further and higher education*, HMSO.

Health and Safety Executive (1990) *Guidance Note GS 48 – Training Standards of Competence for Users of Chain Saws in Agriculture, Arboriculture and Forestry*, HSE.

Health and Safety Executive (1991) *Guidance Note GS 50 – Electrical Safety at Places of Entertainment*, HSE.

Health and Safety Executive (1992) *Management of Health and Safety at Work Approved Code of Practice.*

Health and Safety Executive (1992) *Personal Protective Equipment.*

Health and Safety Executive (1992) *Work Equipment.*

Health and Safety Executive (1992) *Workplace Health Safety and Welfare Approved Code of Practice.*

Health and Safety Executive (1993) *Guidance Note GS 31 – Safe Use of Ladders, Step-ladders and Trestles*, HSE.

Health and Safety Executive (1993) *Guidance Note GS 42 – Tower Scaffolds*, HSE.

Health and Safety Executive (1993) *Step by Step Guide to COSHH Assessment*, HMSO.

ILAM (1995) *Safety in Indoor Adventure Play Areas*, ILAM/NPFA/ROSPA.

Jackson, Albert and Day, David (1991) *Collins Good Wood Handbook.* Harper Collins.

Kids' Clubs Network *Food safety and kids' clubs*, undated.

King, K. and Ball, D. (1989) *A Holistic Approach to Accident and Injury Prevention in Children's Playgrounds*, LSS.

King, R.W. and Hudson, R. (1985) *Construction Hazard and Safety Handbook*, Butterworth.

Mott, A. (1994) *Patterns of Injuries to Children on Public Playgrounds*, Archives of Disease in Childhood, 71.

National Children's Play and Recreation Unit (1992) *Playground Safety Guidelines*, DES and Welsh Office.

NPFA Technical Advisory Notice, TAN09, *NPFA Interpreted BS Dimensions.*

NPFA Technical Advisory Notice, TAN01, *Children's Play Equipment – BS 5696 and NPFA Recommendations on the Extent of Impact Absorbing Surface Area.*

NPFA Technical Advisory Notice – TAN 20 (1993) *Children's Play Equipment – BS 5696 and NPFA Recommendations on Slip Resistant Surfaces for Ladders, Stairs and Ramps.*

NPFA Technical Advisory Notice – TAN 22 (1994) *Notes on the Construction and Maintenance of Children's Sand Pits*.

Potter, D. (1994) *Don't go near the monkeys, children*, District Councils Review, November.

Potter, D. and Heseltine, P. (1989) *Impact Absorbing Surfaces for Children's Play Areas*, Play Wales.

ROSPA (1995) *Constructional Notes for Self-build Play Equipment*, January.

ROSPA *Childhood Accidents*, undated.

ROSPA (1994) *The Children's Playground*.

ROSPA *Playground Design*, undated.

Chapter 6

Materials and equipment

6.1 INTRODUCTION

The quality of children's play experience is linked to the variety of resources available for use in their play, and the way in which those resources are presented and made available. Adventure playgrounds should provide an ever-changing range of materials and equipment as resources for the use of children and to enhance the play value of the indoor and outdoor environments.

The purpose of this chapter is to highlight the health and safety issues concerned with the selection, storage and use of materials and equipment.

The selection of materials for use on adventure playgrounds should first and foremost be made on the basis of their potential to enhance the play experiences of children. However materials used on adventure playgrounds should also be selected with regard to the unacceptable risks inherent in their use or storage.

6.2 HAZARDS

6.2.1 Substances

The Control of Substances Hazardous to Health (COSHH) Regulations, 1994 impose a duty on employers to prevent exposure by employees and others affected by their work to substances hazardous to health, or if prevention is not attainable to adequately control that exposure. A risk assessment process is required which identifies where hazardous substances are present in the workplace and the potential risks associated with them. A five-stage assessment procedure for hazardous materials is described in Part Two of this publication in the section on the COSHH Regulations.

These can be recognized by experience, from information provided by suppliers including product data sheets, and by asking advice. Some Health and Safety Executive Guidance Notes provide specific information on dangerous substances.

At an adventure playground these could include: domestic cleaning materials, paints, solvents, adhesives, pesticides, herbicides, plastic films, dyestuffs, aerosols etc. As a rule of thumb any substance which is labelled 'Keep away from children' must be considered hazardous.

Once identified, it is necessary to establish how the substances can be hazardous. This will require looking at how and where they are stored and used, who may be affected, how they would be affected (e.g. swallowed, breathed in etc.), and what measures are currently being taken to prevent or control exposure. If as a result of this assessment it has been concluded that there is a risk to health, then in order to comply with the regulations it will be necessary to:

- prevent exposure – by changing the process, replacing it with a safer alternative or using it in a safer form;
- or, if prevention is not reasonably practicable, adequately control exposure by a combination of the following – total enclosure of the process, partial enclosure, general ventilation, using systems of work which minimize chances of spills.

These measures should include consideration of ways to ensure that children do not come into contact with hazardous materials. This may include using some materials when children are off the site. Where tasks involving potentially hazardous materials must be undertaken when the site is open, a full assessment of the likely risks to children and playground users must be undertaken.

Where materials are to be purchased a data sheet must be obtained from the manufacturer of the product. This will provide information concerning the potential hazards of the product and information on safe use, handling and storage. The availability of alternatives with no risk, or less risk, attached to their use and storage must be investigated.

Where materials used on adventure playgrounds are scrounged or recycled, staff will often need to undertake a COSHH assessment without information from the manufacturers. This must be completed before the material is used. If there is insufficient information concerning the materials to make a reliable assessment it must be refused, or properly disposed of, as soon as possible.

Where work is necessary to prevent or lessen hazards (for example, the removal of nails from timber) this must be done as soon as possible. The material must be stored securely until the sorting and rendering safe is complete.

Scrap materials used for arts and crafts activities must be assessed for potential hazards and treated in the same way as other materials which do not have product data sheets to accompany them.

Once an assessment has been made it must be recorded, and reviewed regularly.

The Health and Safety Executive or Environmental Health Department of the local authority will be able to give advice on matters concerning the application of COSHH Regulations to specific products.

6.2.2 Equipment and materials

Most things provided as equipment for use on an adventure playground are potentially hazardous in one way or another. For example, most are flammable and will give off noxious fumes if burnt; some are heavy; many have sharp edges or corners; small items can be swallowed and not all equipment is clean or kept clean. If used by children older or younger than those for which the equipment is designed, it may become a hazard.

It is therefore good practice to undertake a risk assessment of each item of equipment. This will identify potential hazards and strategies to minimize those hazards.

Where items of equipment are covered by a British or European Standard they must conform to that standard; for example, the toys must bear the CE mark or the Lion Mark (British Toy Manufacturers' Safety and Quality Symbol).

The Children Act Guidance and Regulation, Volume II requires local authorities to satisfy themselves that equipment provided for children under eight years of age is fit, having regard to its condition, situation, construction and size. The following will be considered in the assessment of fitness of equipment. It should:

- be appropriate to the ages and stages of the children who will use it;
- conform to a British Standard if one exists, or a European Standard if appropriate;
- be adequate in quality and type for the number of children attending the facility.

6.3 STORAGE

The availability of adequate and appropriate storage facilities for all materials and equipment will greatly contribute to lessening the hazard potential of the materials themselves or of their presence on site. Anything which cannot be stored safely must not be kept on site.

All materials delivered to the site must be checked and immediately removed to appropriate storage areas. Materials unsuitable for use by children must be stored in secure storage areas.

All toxic or flammable substances must be locked away from access by children. All such substances (paint, cleaning materials, chemicals for use in craft activities, preservatives etc.) must be stored in clearly labelled original sealed containers. A complete list of items stored must be prominently displayed in the place of storage. The product data sheets and/or summaries of information should be readily available.

6.3.1 Storage arrangements

Arrangements for storage must include consideration of the separate storage of incompatible materials (e.g. corrosive substances and flammable liquids). The

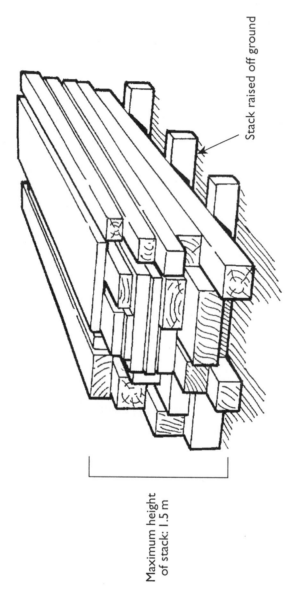

Stack raised off ground

Maximum height
of stack: 1.5 m

As an additional precaution the bundle should be tied with rope or nailed timber

Figure 10. **Stacking timber.**

product safety data should give advice on this matter.

Hazardous or potentially hazardous work equipment must be safely stored in a secure store. Electrical or power tools, and all equipment not suitable for use by children without close supervision must only be accessible to workers or other authorized adults.

Foam-filled equipment must be stored in such a way that if set alight the consequences are as far as possible minimised. This could include provision of a separate storage area with fire-check doors which are kept closed. The storage of this equipment must be considered as part of the fire prevention measures.

Tools and equipment for general use must be stored safely. All blades must be covered. For ease of access, each tool must have its individual position in the store. Where tools are hung on a board or wall, fixings must be secure and appropriate. Tools must be replaced after use.

Special storage areas for items such as rope, cable, ladders and heavy equipment must be suitable for the weight and bulk of the item. There must be adequate, clear access to enable the item to be removed and replaced easily by an appropriate number of people.

Heavy items must not be stored on shelves or other fixtures which are unsuitable, inadequate in construction, or too high to allow safe removal or replacement.

Shelving and other storage fixtures must be sufficiently large for the items to be stored. Items must not project beyond the edge of shelves. Where small or unstable objects are to be stored, they must be contained in boxes, trays or similar containers and appropriately labelled.

Timber and other materials used in the outdoor area of the adventure playground must be stored in a clearly defined area.

Wood intended for particular uses should be kept apart from general resource timber. Separate compounds are recommended. All stored building materials must be kept away from buildings, structures and other activity areas. Materials must not be stacked against, or adjacent to fencing, to minimize opportunities for improvised exit/entry. The use of storage areas for construction or other play activities must be discouraged and children must never be allowed to play on timber stacks. Storage areas must be kept tidy and clear of debris and stacks of timber regularly checked for stability and the condition of the wood.

Timber stored for future use must not be in direct contact with the ground. Care must be taken to protect it from the weather, since rot may set in. Timber must be carefully stacked horizontally, to a maximum height of 1.5 m (see Figure 10).

6.4 TIMBER

It is difficult to specify the precise strength of the timber used in adventure playground structures. In general the use of structures will be very intensive and, therefore, those involved in construction should over-specify all components and fastenings, particularly where recycled materials are used. Timber with annual growth rings close together such as Scots Pine and Norway Spruce should be used. Other softwoods suitable for building construction include larch, European redwood, douglas fir and western hemlock. Suitable hardwoods such as ash, European, Japanese and American oak, European beech, aformosia, Brazilian mahogany and teak are preferable where the timber is subject to hard wear. Hardwoods should be from renewable sources.

Inevitably some timber will be unusable when delivered, or will deteriorate after use. Damaged or scrap timber must be kept separately and the storage area clearly marked to ensure that it is not used for construction. It must not be allowed to accumulate and should be used as firewood, or otherwise safely disposed of. Pressure-treated timber must not be burnt.

It must be made clear to workers and children which materials are appropriate for construction and other purposes. When used wood is collected from the site after use it must be checked before being returned to the appropriate storage area.

It is difficult to predict the useful life of all secondhand/recycled timber. Its age, condition and dimensions must be carefully considered for each individual play structure. Even though superficially it may appear sound, its practical strength may be questionable. Further, children may seek out many ways to amend or undo the structural work of the workers as part of their daily adventurous activity. Their intentions will not usually be malicious. Nonetheless these factors all indicate the need to be particularly careful in the choice of timber and the use to which it will be put. As with all playground equipment and materials, regular inspection and maintenance must take place.

Where timber will come into direct contact with the skin, for example on slides or hand-rails, special care must be taken to ensure safety. Certain timber will splinter easily and this should be sanded down as required and removed if too dangerous. Sharp edges must be rounded off.

All wood to be used in permanent contact with the ground, and any part of such wood up to 300 mm above ground level, should be treated with preservative before it arrives on site. ROSPA advise that the only preservatives suitable for play structures are those derived from copper-chromium-arsenic (CAA) compounds, which are pressure impregnated. Although initially toxic, ROSPA advise that timber treated in this way is chemically fixed after 14 days (but recommend waiting 42 days before using it and that suppliers must be asked to remove any surplus salts before delivery). Timber treated in this way is suitable for ground contact. Some timber is treated with Vacsol organic solvent timber preservatives, through a double vacuum process, although this is not suitable for contact with the ground. Suppliers should be asked to advise on the most appropriate treatment for the use to which the timber is to be put. Treating the timber with Aquaseal will give additional protection.

Suppliers should provide data sheets with the timber, which will include advice and information on handling and disposal of treated timber. Good practice includes the following.

1. Handling
    • If the timber is still moist from the treatment don't handle it and refuse to accept it on site, or if it is already on site keep it away from children and adults.
    • Cuts and abrasions must be protected with waterproof dressings.
    • Wash hands before eating and drinking.

2. Disposal

    • Do not use sawdust as animal litter.
    • Do not burn.
    • Contact a waste disposal authority for advice.

Guidance is available from the Health and Safety Executive on in-situ treatment using timber preservatives (Guidance note GS46). Proper in-situ treatment using preservatives requires staff who are properly trained. A survey should be undertaken by a trained and competent person, pesticides used must be approved under the Control of Pesticides Regulations, and a range of hazard prevention measures must be put into operation. If in-situ treatment of timber is required on an adventure playground, advice and guidance must be sought from the Health and Safety Executive.

Creosote is widely used as a preservative on recycled material used in playgrounds, such as telegraph/electric poles and railway sleepers. It is a potentially hazardous substance as it is toxic, flammable, corrosive and may cause damage to skin or clothing. Risk assessment undertaken at the design stage for structures must identify if and where it is appropriate to use timber which has been treated in this way. Creosote must not be applied as a preservative above ground. The use of secondhand poles for main uprights or structural components is not recommended.

Recycled railway sleepers may be contaminated by sewerage, and often weep tar in hot weather which is noxious, contaminates skin and clothing and is highly flammable.

6.5 ROPE

Generally the choice of rope will be determined by its suitability for a particular purpose. The use of secondhand rope should be avoided.

Wherever rope is used it must be appropriate to the loading and strain to which it will be subjected. The safe working load must always be borne in mind and rope should be capable of bearing several times the anticipated maximum load.

Rope is vulnerable to wear. The use of appropriate fixings and hardware will reduce the rate of deterioration, but will not prevent it entirely. All points of contact between rope and any other material must be suitably protected to reduce friction.

It is also vulnerable to deterioration through the weather or the elements, although most new rope is 'waterproof' or 'rotproof' if it is made from a natural fibre. The prime causes of damage are heat, light and moisture. Natural fibre ropes, such as manilla or sisal, are particularly susceptible to moisture. They must be stored in a dry place, preferably hung up so as to allow air circulation. They will take a long time to dry thoroughly, and unless stored properly, will be subject to rot and mildew. Synthetic fibre ropes, such as nylon or polypropylene, are resistant to rot and the effects of moisture, but vulnerable to stretching or distortion when exposed to heat and light.

Wire rope with a nylon sheaf is widely used in purpose-built play structures. This material is hard wearing and resistant to the effects of the weather. It must be checked for nibs (sharp ends of the wire from which the core is spun) and for non-accidental damage.

All rope must be frequently and regularly inspected for the effects of damage, wear, deterioration and contamination. Inspection of the whole external surface of the rope must be supplemented by examination of the internal condition of the strands. Light ropes can be opened by twisting in the reverse direction to the twist of the strands, while heavier rope may need to be opened by easing the strands apart with a spike. Care must be taken not to damage the fibres.

Rope, especially new rope, can stretch by an additional one-third or more of its original length. Therefore it should be checked hourly on the first day of use, and at least daily for the next week, and adjusted as necessary.

To prevent fraying at the ends, natural fibre ropes may be whipped with twine or back-spliced. Alternatively, the point where the rope is to be cut may be bound with several layers of heavy tape. A cut through the tape will leave both ends bound. Fraying of the ends of synthetic rope may be prevented by applying heat to fuse the fibres. Overheating may cause flame or molten drips, so protective gloves must be worn.

When ropes are to be fixed or joined, only the appropriate standard knots or splices must be used. Ropes, particularly when used in swings or other moving equipment, must not be extended by the addition of further lengths. Ropes of differing thicknesses must not be joined. It must be born in mind that all knots and splices may reduce the strength of the rope.

6.6 WIRE CABLE

Where cable is to be used because of its advantages in strength, durability and resistance to accidental or malicious damage and wear, care must be taken to utilize those advantages by the correct methods of handling, fixing and maintenance.

Cable, when not in use, must be stored in loose coils to reduce the risk of permanent kinks caused by over-bending. It should be oiled to reduce the possibility of rust, and stored in a dry place.

When handling cable, heavy protective gloves and eye protection must be worn. If the coil is heavy, protective boots should be worn. Minor damage or wear can break individual wires producing 'nibs' or splinters which may be hazardous. The end of the cable must always be bound to prevent fraying. When cable is to be cut, the area surrounding the cut should be wrapped in several layers of heavy tape applied in a spiral and the taped area cut through with a hacksaw, leaving both ends bound (see Figure 11). Please also note the following points:

1. Where a fixed loop has not been provided in the cable, all fixings of cables must have the required number of grips, and be correctly fixed. See the section on ariel runways for details.
2. All fixings must be regularly inspected, and maintained and replaced as necessary.
3. Cable must never be used under permanent tension as failure of the cable or its fixings may result in a dangerous whiplash effect.
4. Whenever cable is in contact with wood or other material, it must be suitably protected to reduce friction (see Figure 7).
5. Cable should be regularly greased if it is to used as the running line of an aerial runway or in some other situations where it is subject to friction. Where it is static, it should be oiled regularly to reduce corrosion. For more about cable use in aerial runways see Chapter 5.

6.7 METAL FIXINGS

Bolts or studding with suitable plates or large-diameter washers must be used for fixing all structural members of structures. Coach screws may be used for the fixing of lighter elements or diagonal bracing. Nails or screws of the correct size may be used for planking, boarding and small joists.

All metal fixings should be treated to resist corrosion. The use of galvanized fasteners or treatment on site with a proprietary substance appropriate to the metal being treated, and used in accordance with manufacturers' instructions, is recommended.

Bolts and screws must be countersunk or suitably covered (see Figures 4 and 5). All practical steps must be taken to ensure that components are set flush. When studding is used, the bolt should be secured by means of a lock-nut.

Wire rope, cable or chain used on the playground will generally be subject to friction and stress. All fixings and accessories must be of the correct size, of appropriate design and suitable material. It is important to remember that the rate of metal to metal wear through friction is accelerated where metals of different hardness are used. All fixings and accessories should be of hardened steel, including pulley wheels and spindles. They must be inspected regularly at the points of wear.

It is likely that components used on playgrounds will be subject to more intensive use and thus greater wear than in their normal application. The greatest care must be taken in their selection. Where reference is made to a British Standard, consideration must be given to any difference in application, and specialist advice sought where necessary. PLAYLINK can advise on these matters.

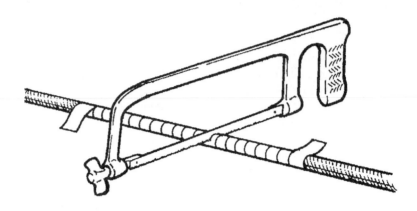

Figure 11. Cutting cable.

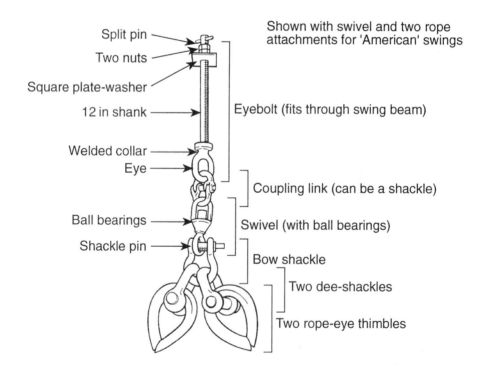

Figure 12. Example of eyebolt fitting.

All metal components must be examined regularly and often and replaced as soon as wear becomes evident. In the case of pulley blocks or shackles it is sufficient to replace the worn component(s). Wherever possible, steps must be taken to minimize the effects of friction – for example by the use of thimbles inside loops, by sleeving or padding at potential rubbing points, or by the use of alternative fixing methods. The use of improvised fixings must be avoided, as the adapted component may be unable to cope with stresses different to those for which it is designed.

Eyebolts

These are used for terminal anchoring for rope. The eyes must always be cast or welded closed. The eyebolt must be secured with a lock-nut and washers and plates used to spread the load. Where eyebolts pass through timber, the timber must be treated against rot and the bolt against corrosion. Regular inspection of the shank of the bolt and of the timber must be undertaken (see Figure 12).

Shackles

These are used on cable, rope and chain when they are to be joined or fixed to an anchoring point or another length of cable, rope or chain. For extra safety or to prevent tampering, the removable bolt should be secured with a grub-screw, split pin or similar, or replaced with a hardened steel bolt of sufficient length to accept a lock-nut (see Figure 13).

Bulldog grips/cable clamps

These are used to hold a loop in the end of wire cable and they should conform to DIN 1142. Great care must be taken to apply these grips/clamps in the correct manner – the U-bolt on the untensioned/unloaded end of the cable and the grip itself on the standing part. The number of grips required relates to the diameter of the cable (see Chapter 5 and Figure 14).

Thimbles

These are used to protect the inner surfaces of a loop in rope or wire cable. It is important that the correct size of thimble for the rope or cable is used. Thimbles must be securely held within the loop – in wire cable by the use of bulldog grips, in rope by the use of whipping to tighten an eye splice (see Figures 9 and 14).

Turnbuckles/bottle screws/rigging screws

These are used to take up slack on cables. Turnbuckles used as permanent fittings must be as strong as possible. Only closed-eye turnbuckles must ever be used – the closure of open hooks by 'mousing' is not acceptable. For safety and to

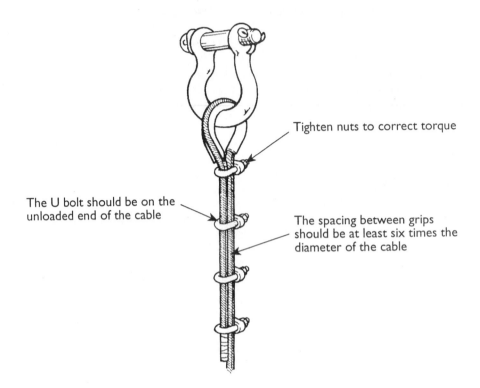

Tighten nuts to correct torque

The U bolt should be on the
unloaded end of the cable

The spacing between grips
should be at least six times the
diameter of the cable

Figure 13. Shackle.

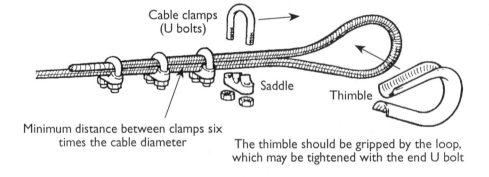

Cable clamps
(U bolts)

Saddle

Thimble

Minimum distance between clamps six
times the cable diameter

The thimble should be gripped by the loop,
which may be tightened with the end U bolt

At least four clamps should be used; more
may be necessary. See text for details

Figure 14. Cable fixings.

prevent tampering, turnbuckles must be wired between the eyes so that the potential drop is no more than 50 mm.

Purpose-made fittings

These are now readily available for purchase from manufacturers of playground equipment. These components are useful, as they are designed and constructed for similar purposes for which they are required on adventure playground structures. The manufacturers' advice should be sought on compatibility of uses. Purchasers must ask for written confirmation that items have been tested in accordance with the methods described in BS 5696 Part 1 to the standards set out in BS 5696 Part II.

6.8 TYRES

The following points should be noted.

1. Tyres have many uses on playgrounds but they present a fire hazard and only those in use and those properly stored as spares should be kept on site.
2. Most tyres have steel bracing in them and tyres must not be used when damaged strands of wire protrude.
3. Tyres kept for used in fixed positions on the playground should have holes drilled to allow water to drain away.
4. Tyres that are fixed together, eg to form a climbing 'net', must not form a wedge trap which is above 1 m from the ground. See Chapter 5.
5. Informal use of tyres should be supervised. Care must be exercised to minimize accidents caused by tripping or being knocked over by rolling tyres.
6. Only nylon braced tyres must be used as swing seats, where their impact absorbing qualities are important. They also have no potentially dangerous wires in the tread or sidewalls.

6.9 PAINT

It is probable that adventure playgrounds will use quantities of paint. Structures, buildings and fencing may be painted regularly. Selection, storage and use of paint will be considered within the risk assessment required under COSHH Regulations.

Household paint is often readily available in small quantities left over from decorating and given to the playground. All secondhand tins should be opened on receipt and inspected and if suitable, resealed and placed in a secure store. Large quantities should not be held, as they represent a serious hazard in case of fire. Empty tins, or those not accepted for use, must be disposed of safely – preferably by immediate removal from the site.

It is preferable to use emulsion or acrylic (water-based) rather than gloss (oil-based) paints for activities involving children, as most emulsion and acrylic is comparatively easy to remove from skin and clothes.

The use of white spirit, turpentine or other highly flammable cleaning material should be avoided wherever possible. Proprietary products may be suitable. Many varieties of paint are available which require only water and mild detergent to clean brushes and skin. Where spirit-based solvents are used they must be considered within the COSHH risk assessment. They must be stored only in small quantities, in a secure place and, in accordance with manufacturers' instructions. The use of chemical paint strippers must be avoided.

Paint with high lead content must not be used in adventure playgrounds. Most domestic brands have low levels of lead, but many 'trade' paints and primers have unacceptably high lead content. If in doubt, inquiries must be made of the manufacturer and/or the local Trading Standards/Consumer Protection agency.

6.10 CRAFT MATERIALS

Most materials used in craft activities are not intrinsically hazardous. However most are flammable and some give off noxious fumes. Some are very heavy, either individually or in bulk. Many have hard edges or sharp corners. Small items can be swallowed, or stuck in the throat, ears nose etc. Most items are not clean. The following list highlights hazards associated with scrap material as identified by Cardiff Play Resources Centre:

- paper and card – flammable, sharp edges, not necessarily clean, the paste on some ready-pasted wallpapers contains fungicide;
- plastic – flammable, not necessarily clean, sharp edges and corners which may shatter, can be very heavy, polythene bags can suffocate, never use small pieces with children under five years of age;
- fabric – often flammable, not necessarily clean, often not suitable for clothing;
- dyestuffs may be toxic;
- rope, string, wire, cotton – flammable, can cause cuts and friction burns;
- wood – flammable, heavy, can splinter, may be treated with chemicals;
- foam/sponge – can be highly flammable;
- metal – heavy, sharp edges and corners.

The use of the following is not recommended:

- rubber – flammable, dirty, black rubber contains carbon which is toxic;
- ceramics – heavy, break easily and may splinter leaving sharp edges and corners;
- glass – shatters, sharp edges;
- expanded polystyrene – flammable, toxic when burnt, when cut can produce small fragments which can block nasal passages and cling to the walls of lungs and does not show up on X-rays.

This information should not be treated as complete, and the hazards of all materials must be identified within the general risk assessment, which should include consideration of the age of the children who will be using the materials.

They must be stored safely and where appropriate, their availability to children restricted by the use of small containers. The use of glass or other breakable containers for paint, brushes or water must be avoided.

Where craft activities require the use of chemicals (for example, dyestuffs or photography or pottery) these must be stored securely and used only under the supervision of an experienced, competent adult.

Where there is a specific area set aside for art and craft activities, every effort must be made to keep it clean and to ensure that is free form hazards. Work surfaces and flooring must be designed to minimize the risk of accidents following spillages.

6.11 SOFT PLAY AND INFLATABLES

The use of soft play equipment and, to a lesser extent, ball pools has become a common feature on many adventure playgrounds. Advice and guidance on the purchase, installation, use and maintenance of these items is provided in the publication *Safety in Indoor Adventure Play Areas*. (See Further reading, below.)

Guidance for the safe use of inflatable play structures is contained within the Health and Safety Executive Guidance Note PM 76, *Safe Operation of Passenger-carrying Amusement Devices – Inflatable Bouncing Devices*, which describes various factors that can contribute to accidents on this equipment and precautions that should be taken to avoid them. This Guidance Note should form part of the safety information records held on an adventure playground where inflatables are used. Its main recommendations are concerned with:

- safe access;
- safe operation, including supervision levels;
- training of attendants;
- examination, inspection and maintenance, (including the electrical components);
- design and manufacture.

Inflatables are often very heavy and cumbersome to lift and attention must also be given to safe arrangements for handling and transportation.

6.12 UNSUITABLE MATERIALS

If any doubt exists about the suitability of any material, and information on labels is not sufficient to assess suitability, expert advice must be sought. Where no firm conclusion can be drawn, the material must not be used. Substances in unmarked containers must not be used.

In many cases, non-toxic and/or non-flammable alternatives are available and should be used instead. For example:

- water based, acrylic (emulsion) instead of oil-based (gloss) paints;

- water-based, non-flammable contact adhesives instead of solvent-based contact adhesives.

Where no suitable alternative is available, toxic and flammable materials must be kept to a minimum and must be used in the context of a risk assessment.

There are many materials which may be readily available for use in structure building but which are unsuitable, such as:

- corrugated iron, which is likely to rust and deteriorate, and in any case, has sharp edges;
- polythene sheeting, which is a fire hazard;
- chipboard, block board and most other laminates, which have low inherent strength and deteriorate immediately when wet;
- scaffold poles, which are unlikely to possess sufficient strength and need expert fixing.

It is essential that playworkers considering the use of any unorthodox materials in structure building must first satisfy themselves as to their safety and suitability and must seek specialist advice where necessary.

Asbestos cement (in sheeting, piping, guttering, insulation or woven fabrication form) is highly toxic, explodes in fire and must never be used on playgrounds.

6.13 LEGISLATIVE BACKGROUND

Part Two of this publication contains summaries of the following legislation referred to in this Chapter:

- The Health and Safety at Work Act (1974) and its Regulations:
  - Control of Substances Hazardous to Health (COSHH) Regulations 1994;
  - Management of Health and Safety at Work Regulations 1992;
  - Fire Precautions Act 1971, amended by the Fire Safety and Safety of Places of Sport Act 1987;

- The Safety Signs Regulations 1980;
- Food and Environmental Protection Act, 1989 Part III –The Control of Pesticides Regulations
- Environmental Protection Act 1990.

6.14 FURTHER READING

The following are the main source documents for this chapter:

Cardiff Play Resource Centre *Safety of Scrap Materials Information Sheet*, undated.

HAPA Information Sheet 5, *Play Structures and Play Equipment*, undated.

Institute of Ceramics (1991) *Health and Safety in Ceramics*, 3rd edition.

Health and Safety Executive (1992) *Management of Health and Safety at Work Approved Code of Practice*.

Health and Safety Executive (1993) *Step by Step Guide to COSHH Assessment*, HMSO.

Health and Safety Executive (1992) *Work Equipment*.

Health and Safety Executive Guidance Note PM 76 (1991) *Safe Operation of Passenger-carrying Amusement Devices – Inflatable Bouncing Devices*.

ILAM (1995) *Safety on Indoor Adventure Play Areas*, ILAM/NPFA/ROSPA.

The Children Act Guidance and Regulations, Volume II – *Family Support, Day Care and Education Provision for Young Children*, HMSO, 1991.

The Children Act Guidance and Regulations, Volume 8 – *Private Fostering and Miscellaneous*, HMSO, 1991.

Chapter 7

Tools and work equipment

7.1 INTRODUCTION

In an adventure playground, staff, volunteers and children will all be participants in the construction and maintenance of facilities, and in doing so will use tools and work equipment. It can be satisfying and enjoyable for all concerned, but raises issues of health and safety which reach beyond those in conventional working environments.

Decisions concerning the safe use of tools must be based on the requirements of the Provision and Use of Work Equipment Regulations 1992 which requires employers to ensure the provision of safe working equipment.

Other Regulations which are relevant are the Electricity at Work Regulations 1989 as interpreted by the Health and Safety Executive Guidance Note GS 23 (Electrical Safety in Schools), which includes regulations and guidance on the provision of electrical tools. The Manual Handling Operations Regulations 1992 and the Personal Protective Equipment at Work Regulations 1992 include provisions relating to tools and work equipment.

7.2 SELECTION OF TOOLS AND WORK EQUIPMENT

Suitable and appropriate tools and work equipment must be provided on adventure playgrounds. The actual tools and work equipment chosen for an adventure playground will depend upon the range of activities undertaken. Employers are required to ensure that work equipment is constructed or adapted to be suitable for the purpose for which it is to be provided. This requires an assessment of:

- the initial condition of the equipment;
- the place where it is to be used;
- the purpose for which it is to be used.

On an adventure playground the abilities of the users should also be considered when selecting tools and work equipment, for example those which are to be used by children.

The assessment will also identify where, as a last resort, personal protective equipment is required to avoid or limit risks associated with the job and/or tools and work equipment used, for example eye, protection while using an electric drill.

Some tools and work equipment may not be suitable for all the groups of users and some may only be suitable for use under supervision. These can be chosen for a particular group if their use can be appropriately regulated through use of a colour coding system or otherwise. Certain work equipment, such as power or electrical tools, must not be used by children. Children must be instructed in the correct use of tools.

The acquisition of specialist tools and work equipment must be carefully examined and consideration given to hiring rather than purchasing items which are not needed frequently.

The temptation to economise by purchasing cheap items must be avoided, as badly made or unsuitable tools are potentially an unacceptable risk.

A list of basic tools is provided at Appendix 19.

7.3 USE OF TOOLS AND WORK EQUIPMENT

Risk assessment will have identified the tools and work equipment suitable for a particular job. Playworkers must be competent and confident in their use. Employers are required to provide employees with adequate information and training in the use of tools and, where appropriate, written instructions. Where people other than employees use tools they must also be provided with appropriate training and information in a way and at a level which is accessible to them.

The risk assessment will also have identified tools and work equipment which is not suitable for all groups of users, and the need for restrictions on when and where tools must be used. For example, certain tools are appropriate for use in a workshop but not for general use in the playground.

The issue and the use of such tools and work equipment must be properly regulated in accordance with health and safety procedures.

Tools and work equipment must be marked if instructions and/or warnings apply to their use. These markings must be easily perceived, understood and unambiguous.

All tools and work equipment must be used and stored in accordance with the manufacturers' recommendations.

7.3.1 The use by children of tools and work equipment

Children should be encouraged to contribute to the development and maintenance of the adventure playground and in doing so will use tools and work equipment. Risk assessment will have identified which tools and work equipment are suitable for use by children and the conditions for safe use by them. There should be as wide a range of suitable tools as possible, available for children of all ages and levels of ability.

Children must be given information and training which will enable them to learn and identify the utility and possible hazards, of tools and work equipment available to them.

They must be instructed in the correct use of tools, and encouraged to clean and maintain them – but their maintenance work must be monitored by staff. Children must learn the habit of holding tools firmly by the handle. When using saws or sharp-edged tools, the cutting edge must be aimed away from the hands and body. Hands must not be placed in front of the cutting edge or around it. In order to make supervision easier, children should work in small groups. When using riskier tools (such as spades and shovels) children will need closer supervision. Children must never be given tools which they are unable to handle safely. Any tools likely to be hazardous when used by children must never be left unattended.

Where personal protective equipment is provided, as a last resort to avoid or limit hazards when tools or work equipment are used, it should be the correct size and appropriate fitting for the children using the tools and/or work equipment.

7.3.2 Inspection and maintenance

The Provision and Use of Work Equipment Regulations 1992 place an obligation on employers to ensure that work equipment is maintained in an efficient state, working order and good repair. 'Efficient' in this context relates to how the condition of the equipment might affect health and safety.

Maintenance can be:

- routine – which includes periodic lubrication, inspection and testing based on the recommendations of the manufactures;
- planned preventative – which may be necessary where inadequate maintenance could cause the equipment to fail in a dangerous way.

It is recommended that a record of maintenance should be kept.

Routine maintenance for non-electrical tools could include:

1. *handles* – any tool with a loose handle is potentially dangerous, as control over the tool is reduced and there is an unacceptable risk that the working part of the tool may work free and fly off; where handles have been damaged the tool should be withdrawn from use until repaired
2. *cutting edges* – should be checked for wear or damage as a blunt tool is not only less effective but it is also more likely to slip causing damage to the work or potentially serious injury to the user;
3. *spades and shovels* – must be checked for damage to the blade and handle and the cutting edge of a spade should be resharpened periodically;
4. *regular cleaning*;
5. *metal parts* – especially blades of cutting and digging tools must periodically be lightly oiled. This will help to protect the surface from corrosion;

6. *screws, nuts and other fastenings* – must be regularly checked and tightened;
7. *new blades* – must be regularly fitted where appropriate and old blades disposed of safely;
8. *chisels, planes, drills and auger bits* – must regularly be sharpened and saw teeth set, in accordance with manufacturers' or suppliers' instructions and using only the correct equipment and techniques;
9. *moving parts* – must be lubricated in accordance with the manufacturers' instructions.

The maintenance of tools is the responsibility of the staff, but it is advisable that replacement of handles, hammers and all other swinging and digging tools be undertaken by specialists.

Guidance on the inspection and maintenance of ladders and step ladders is contained within the Health and Safety Executive publication, *The Safe Use of Ladders, Step Ladders and Trestles*, GS 31 (reprinted July 1993). These are:

• they must be individually marked and inspected before and after normal use and regularly by a person competent to do so (a person who has practical and theoretical knowledge and actual experience of ladders);
• climbing and gripping surfaces must be free from oil, grease or mud;
• timber items must be checked for rot, decay or mechanical damage and rungs must be checked for looseness;
• metal ladders must be checked for twisting, distortion, oxidization, corrosion and excessive wear;
• other checks are for cracked or broken rungs or stiles, rivets, tie rods, hinges etc.

Electrical work equipment and tools are subject to the requirements of the Electricity at Work Regulations 1989. It recommends in respect of the types of tools and equipment on adventure playgrounds:

1. the compiling of an inventory of equipment and associated maintenance and inspection log;
2. that all portable equipment is routinely inspected;

   • class 1 items (which are earthed through the lead to the ringmain), a detailed inspection at least every 12 months, and a test by a competent person including an earth continuity test, a checklist for inspections is included at Appendix 8,
   • class 2 items (which are double insulated and therefore do not have an earth lead to the ringmain) must be subject to the same test routine as for class 1 items with the addition of a visual check before use.

3. all flexible cables (including extension leads) to be selected, maintained and used so that there is adequate protection against foreseeable mechanical damage.

Health and Safety Executive Guidance Note PM 32 gives full details of the safe use of portable electrical equipment.

The use of markings will be helpful in the identification of class 1 and class 2 items. These markings must be easily perceived and understood and unambiguous.

All tools require continual attention to maintain them in good condition.

7.3.3 Storage

Tools must be securely stored in a clean dry place. Every tool should have a regular, clearly marked place in whatever storage system is used. Tools should be marked to enable ease of identification, for example green for tools which can be used by children, red for those to be used only by staff and blue for those which require specialist training, yellow for those requiring close supervision when used by children. Some users will be colour-blind and codes can be supplemented with visual codes such as shapes. They must be checked-in after use. Sharp edges must be protectively covered when not in use. All power tools must be stored under lock and key.

Ladders and step ladders should be stored on racks designed for the purpose, which are easily accessible. Wooden ladders should not be stored by hanging. They should be supported flat. This is necessary to ensure that they lie true.

7.3.4 Carrying tools

It is essential that tools are carried safely, for example chisels must be carried with their blades downwards. Tools must be passed by their handles. When working at height, tools must be passed to and from the worker where possible, otherwise they must be carried in a bucket or hold-all sling. Children must be encouraged to carry all tools carefully, and not to overload themselves. They must not run or take part in active play while carrying tools.

7.3.5 Clothing

Personal protective equipment may be required, as a last resort, in order to avoid or limit hazards when using tools. It must be of the correct size and appropriate fitting. This could include steel-toed boots, goggles, ear-muffs, protective head-gear and/or gloves. See Chapter 5.

Care must be taken to ensure that clothing does not interfere with the safe use of tools. Loose cuffs, overalls/trousers which are too long and buttons missing from overalls/shirts may cause material to interfere in the work. Sharp tools must never be carried in pockets.

7.4 TYPES OF TOOLS

7.4.1 Power and electrical tools

Electrical tools are subject to specific requirements of the Electricity at Work Regulations 1989 (see above). Electrical and other power tools must only be used by workers competent in their use and strictly in accordance with manufacturers' instructions. Work with power tools is best undertaken when children are off the site and must never be undertaken when the worker is alone on site. They must never be left unattended and while not in use must be secured to prevent accidental start-up.

Where possible electrical tools must be cordless. If this is not possible they should operate at 110 volts and be used with a 110 volt transformer. A residual current device installed close to the point of supply is recommended for all uses and is required for work outdoors. Plugs and sockets used outdoors must comply with BS 4343 (male/female fittings). Electrical tools must not be used in damp or wet conditions, or in the presence of flammable vapours or gases.

Extension cables must be used with care and only when no alternative supply is accessible. They must not be subject to tension and must be supported along their whole length. They must not be allowed to snag on corners or protruding edges. Where their path crosses pathways or movement flows they must be protected and marked to avoid damage to the cable or accidents due to tripping. They must be fully unwound from reels, in order to avoid overheating as current passes through them. After use, and while being moved to a new working area, they must be carefully wound on a suitable reel. Knots, kinks, cuts and abrasions must be avoided and rectified immediately.

All connections within plugs and sockets must be checked regularly. The external insulation must be stripped to the minimum distance required, and cable grips fitted in plugs and sockets always used. All connections must be made safe with standard connectors (see requirements for plugs and sockets used outdoors, above). Improvised connections or repairs to damaged insulation must not be made.

7.4.2 Saws

All saws must be kept clean and sharp. The blade must be securely and tightly attached to the handle. The blade must be straight and the teeth undamaged. Many saws are sold with plastic teeth-protectors and it is advisable, to protect the saw and for greater safety, to retain these to cover the teeth when the saw is not in use.

The blade should be lightly oiled periodically to avoid rusting. Wooden handles should be lightly coated with linseed oil occasionally to maintain the condition of the wood. When handles are cracked they must be replaced. Sharpening and setting of teeth should only be undertaken by a competent trained person.

Saws in general use will fall into the following categories (see Figure 15).

1. *Rip or half-rip saws*: (4.5 points per inch) – used for cutting along the grain. These are generally not necessary for playground use, where the hand-saw can be used instead.
2. *Hand or crosscut saw*: (6–8 points per inch) – primarily for cutting across the grain but can be used along the grain.
3. *Panel saw*: (10–12 points per inch) – useful for plywood or other sheet/composite timber or for small sections of wood.
4. *Bow-saw/log saw* – with blade in light tension on tubular frame. The bow saw is generally used for cutting heavier timber. The blade is held by a bolt and wing-nut at each end which should be checked regularly.
5. *Back or tenon saw* – these are generally finer than handsaws and are stiffened by reinforcement along the back of the blade. They are often easier for children to use as they are shorter, but they are easy to damage if used without care
6. *Hacksaw* – for cutting metal only. Hacksaws are very fine-toothed with thin blades which are vulnerable to damage and often break. Hacksaw blades must only be used in the hacksaw – the practice of improvising handles is to be discouraged.
7. *Multi-purpose saw* – for general and routine timber cutting many have replacement blades and are useful for den building etc.

7.4.3 Screwdrivers

Good-quality screwdrivers with a hardened blade should be used. The correct size should be used, with the blade closely fitting the slot. This will minimize damage to the screw and to the tool. Phillips, Posidrive and other proprietary patterns of slots must be driven with the correct pattern of screwdriver. The longer the screwdriver the greater the force which can be applied. Care must be taken in the use of ratchet screwdrivers which should be well maintained. 'Yankee' screwdrivers, where the operation of the ratchet releases a spring-loaded helical shaft are potentially very dangerous, and must not be used on the playground.

The blades of screwdrivers should periodically be filed level to remove damage and maintain efficiency.

Electric screwdrivers require special consideration. Only Posidrive, Phillips or crosshead type screws and the appropriate drivers should be used, as slothead screws and drivers are difficult to control. Rechargeable battery-operated electric drills are suitable for most light tasks. Screwdriver attachments for electric drills must only be used on drills which are designed for their use, with specific speed and torque settings for screwdriving. The manufacturers' instructions must be strictly followed, when using electrical screwdrivers. Holes of the correct size and depth for the screw should be pre-drilled. No attempt must be made to drive screws into undrilled materials.

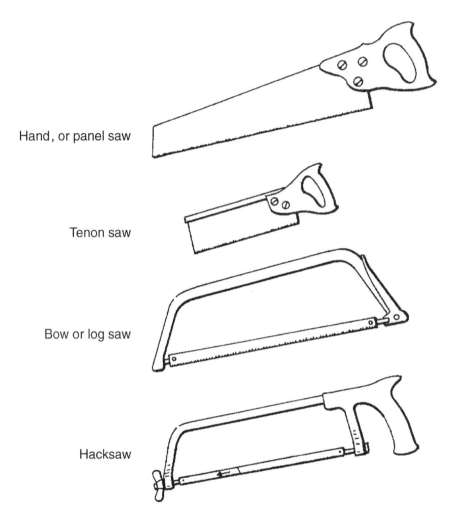

Hand, or panel saw

Tenon saw

Bow or log saw

Hacksaw

A full list of recommended tools is provided
in Appendix 19

Figure 15. Types of saws.

7.4.4 Chisels and cutting tools

All cutting tools need constant and close supervision, efficient maintenance and safe storage. All cutting tools must be regularly checked, sharpened and generally maintained. Old blades must be disposed of safely. Only the appropriate replacement blade or handle must be used.

Chisels must be stored with the edge covered. An individual cover for each blade, removed only while the chisel is in use and replaced immediately afterwards, is most effective. All users of chisels must be instructed in safe techniques for carrying and using chisels. They must be carried blade down one's side – never in a clenched fist or a pocket. No one carrying a chisel must run. Chisels must be kept clean and sharp – blunt or damaged blades are less effective and less safe. Chisels must be sharpened on a whetstone maintaining a regular bevel at a shallow angle, and keeping the face of the chisel flat. They must periodically be lightly oiled to prevent corrosion. The blade must be secure in the handle.

This tool should be used across or with the grain of the wood only. Working against the grain may cause the wood to split or the blade to jump. The edges of the area to be chiselled away must be first cut with a saw (or with the chisel for a countersunk hole not open to the edge of the wood).

The chisel should be hit with a wooden mallet only. The wood to be chiselled must be held firmly. At all times, the blade must be pointed away from the body and both hands must be behind the blade.

Cold-chisels used for cutting metal should be kept reasonably sharp, and the heads occasionally filed to remove burrs and present a good surface to the hammer. Goggles should be used because of the risk of flying splinters of metal. Other personal protective clothing may be required and will be identified in the risk assessment.

Planes need regular maintenance of the sole and the blade. The blade must be kept clean and sharp, and the sole occasionally checked for damage or burrs. Handles must be secure, and the wedge and adjusting screws tight. Planes must only be used on wood which is firmly held. They are primarily workshop tools and instruction and supervision must be given. Portable electric planes should be avoided. Orbital or belt sanders can be used, with special attention given to the manufacturers' instructions for safe operation.

Craft cutters such as lino or wood carving tools should be stored, maintained and used in the same way as chisels. They are potentially an unacceptable risk and supervision and instruction in their use are required. Modelling or general-purpose knives such as Stanley knives with replaceable blades are very sharp. Where these are used, the type with a retractable blade may be more appropriate. Constant supervision is required.

Metalwork files and woodwork rasps must only be used with securely fitted handles. They must be stored separately. The use of rasps for metalwork must not be allowed.

7.4.5 Digging tools

Spades, forks, mattocks, entrenching tools and augers must all be cleaned and maintained regularly. Handles must be checked for damage and replaced with the appropriate handle for the tool. The fixing of the handle to the blade must be checked regularly and any looseness rectified. Blades of spades and augers must be sharpened periodically. Care must be taken in the use of digging tools, children should be away from the excavation. The appropriate footwear must be worn and other personal protective equipment may be required and will be identified in the risk assessment.

7.4.6 Sledge hammers and pickaxes

When using these swinging tools, all observers must be kept well clear. No obstruction must impede the swing of the tool and no one should stand directly in line with the swing either in front or behind the operator. If shafts split or crack, or the head becomes loose, the shaft must be replaced. This is a specialist process and must only be undertaken by a competent, trained person. Sledge hammers and pickaxes must be kept clean. A light coating of oil will help to prevent rust.

7.4.7 Ladders, steps and towers

Safe Use of Ladders, Step Ladders and Trestles, GS 31 (reprinted July 1993) gives basic safety information for users of this equipment. This is discussed in detail in Chapter 5.

Improvised ladders must not be used. Wooden ladders must not be stored by hanging. They should be supported flat. This is necessary to ensure that they lie true. They must always be used on a level base for maximum stability.

7.4.8 Use of tower scaffolds

Scaffold towers are the safest way of working at height particularly for prolonged periods. They should be used only where they can be erected safely to an appropriate height and on a level base. They must be dismantled and stored when not in use.

The Health and Safety Executive Guidance Note, GS 42 (reprinted March 1993) provides useful guidance in the use of tower scaffolds. These are summarized in Chapter 5. The following are the main points which relate to their use.

1. Different types are erected in different ways and have different capabilities and capacities and manufacturers' instructions must be followed.
2. Persons who erect them must be competent to do so, taking into account the complexity of the layout of the tower.

3. Check and follow the loading capacity as specified by the manufacturer.
4. Ideally they must be erected on level, firm surfaces, and in any event the surface must be capable of sustaining the total load. Static towers must have metal base-plates and timber sole-plates.
5. They must be vertical.
6. All towers must be braced in all three dimensions with braces and outriggers, according to the manufacturers' instructions.
7. Working platforms must be at least 600 mm wide and must be properly constructed and fixed.
8. Guard rails 910 mm to 1150 mm high are required on all working platforms from which a person can fall more than 2 m. Toe boards are also required.
9. Platforms must have a safe means of access, always on the narrowest side of the tower, either by vertical ladders internal to the narrow side, internal stairways or inclined ladders, or ladder sections integral with the frame members. Ladders must be lashed to the tower.
10. Unless access is by an inclined stairway, tools and heavy loads must not be carried but should be hauled up within the confines of the tower.
11. Ladders must not be used to extend the height of a tower.
12. Where the tower is used for jobs such as drilling, where considerable sideways force may result, great care must be taken to ensure that the tower is not overturned and the manufacturers' instruction must be followed.
13. They must be dismantled carefully.

The operators of adventure playgrounds where scaffold towers are used are urged to purchase GS 42 from the Health and Safety Executive in order to obtain full details of good practice in their use.

7.4.9 Harnesses

Safety harnesses should be used where there is any risk of injury from falling, and particularly at heights over 2 m. A range of harnesses and corresponding safety aids are available from specialist suppliers and hirers, who will be able to advise on the equipment needed for particular tasks. Anyone who is required to use such equipment must be properly trained to do so. Some suppliers will require proof of such training before purchase/hire, and some will provide training. Harnesses and any other fall-arrest equipment used must meet the performance requirements of BS 1397 and must be inspected by a person competent to do so, and a record of the inspections kept.

7.4.10 Working at height – general

When working at height, care must be taken in the use of tools and in their transport to the work. It is advisable to carry or lift tools and materials in a bag or bucket, which may be fastened at the work-site. The area below the work-site must be kept clear, to avoid accidents caused by dropped tools or material. It is dangerous to over-extend when working at height. If comfortable reach is not possible, then the ladder, steps or platform must be moved to a better position.

7.5 LEGISLATIVE BACKGROUND

Part Two of this publication contains summaries of the following legislation referred to in this chapter:

* The Health and Safety at Work Act (1974), and its Regulations:

    * The Electricity at Work Regulations 1989
    * Manual Handling Operations Regulations 1992
    * Personal Protective Equipment at Work Regulations 1992
    * Management of Health and Safety at Work Regulations 1992
    * Provision and use of Work Equipment Regulations 1992.

7.6 FURTHER READING

The following are the main source documents for this chapter:

Health and Safety Executive (1989) *Guidance Note GS 23 – Electrical safety in schools (Electricity at Work Regulations 1989)*, HMSO.

Health and Safety Executive (1990) *Guidance Note GS 48 – Training Standards of Competence for Users of Chain Saws in Agriculture, Arboriculture and Forestry*, HSE.

Health and Safety Executive (1991) *Guidance Note GS 50 – Electrical safety at places of entertainment*, HSE.

Health and Safety Executive (1992) *Management of Health and Safety at Work Approved Code of Practice*.

Health and Safety Executive (1992) *Manual Handling*.

Health and Safety Executive (1992) *Personal Protective Equipment*.

Health and Safety Executive (1992) *Work Equipment*.

Health and Safety Executive (1992) *Workplace Health Safety and Welfare Approved Code of Practice*.

Health and Safety Executive (1993) *Guidance Note GS 31 – Safe Use of Ladders, Step Ladders and Trestles*, HSE.

Health and Safety Executive (1993) *Guidance Note GS 42 – Tower Scaffolds*, HSE.

Chapter 8

Health and hygiene

8.1 INTRODUCTION

The Health and Safety at Work etc. Act 1974 sets out a general duty on employers and employees not only to be aware of the need for health (and safety) but to ensure that, as far as is reasonably practicable, optimum conditions exist for minimising the risk to health of all those affected by their work. The Workplaces (Health and Safety) Regulations 1992 apply to workplaces which are used for the first time after 31 December 1992, and to existing workplace on 1 January 1996. These Regulations set down requirements in respect of health and hygiene.

The Children Act 1989 and the other legislation listed above also sets out regulation and guidance in respect of health and hygiene.

The site (outdoor area of the adventure playground) is subject to the Health and Safety Regulations, just as are premises and indoor areas.

The general risk assessment undertaken as required by the Management of Health and Safety at Work Regulations 1992 will identify risks to health and issues concerning hygiene.

8.2 BUILDINGS

8.2.1 Cleaning

It is a requirement of the Workplace (Health and Safety and Welfare) Regulations 1992 that furniture, furnishings, floor surfaces, walls and ceilings are kept clean. The Children Act Guidance and Regulation sets out standards for the health and hygiene of buildings, and the general cleanliness of premises will be considered as part of the Registration process.

All areas of the building must be cleaned regularly in a way which is suitable and which does not endanger health. Suitability should be determined through a risk assessment procedure which will include consideration of the requirements of the Control of Substances Hazardous to Health (COSHH) Regulations 1994.

COSHH regulations apply specifically to the use of cleaning materials and the personal protective clothing necessary for users. In practice this means:

- they must be used strictly in accordance with manufacturers' instructions on use, storage and disposal of container and contents;
- cleaning agents must, wherever possible be stored in their original containers, clearly marked with their contents;
- special care must be taken not to use combinations of substances which may be dangerous – for example, some toilet cleaners combine with bleach to produce toxic chlorine gas.

Adequate supplies of cleaning and disinfecting materials must be maintained. The level of stock required will vary according to the size and type of building and degree of use, but all playgrounds should have a stock of all the basic tools and materials, including brooms, dustpan and brush, sponges, cloths, toilet brushes, mop bucket, squeegee, tea towels, hand towels, toilet paper, soap, detergent, disinfectant, toilet cleaner, scouring powder, washing up liquid, surface cleaners and non-slip floor polish.

All cleaning agents must be stored securely, away from children. COSHH Regulations will apply to storage of hazardous materials.

Where cleaning/clearing requires manual handling which may be a risk to employees, the Manual Handling Operations Regulations, 1992 will apply. These Regulations require the provision of appropriate systems of work to reduce the risk of injury as far as is reasonably practicable.

8.2.2 Construction and maintenance

Details of the construction and maintenance of buildings are given in Chapter 4. The following are of importance in respect of health and hygiene.

Areas receiving intensive use must be designed and furnished to facilitate effective cleaning. Particular attention must be paid to floor surfaces and wall finishes.

Heavy-duty, washable flooring surfaces should be used wherever practical. They should be laid so as to avoid the possibility that joins or edges attract or collect dirt. They must never be polished to create a slippery surface – only matt finish polish must be used. Where no additional flooring material is laid, the surface must be treated to make it easier to clean. Concrete floors should be sealed and the sealing process periodically repeated. Wooden floors should be sanded and sealed, and where possible, cracks filled to avoid accumulation of dirt.

Walls should be maintained so as to avoid the problems of dampness or rot, excessive condensation etc. Attention must be paid to wall coverings and decoration. Gloss paint makes a surface which is easy to clean but has disadvantages in terms of expense, ease of application and the need to rub down when redecorating. Other types of paint are also washable and relatively durable without these disadvantages.

Where walls are painted, they should be cleaned regularly (according to use) and repainted periodically. Areas of the wall may, because of the activities carried on near them, require special finishes. Tiled or plastic laminate, or metal splashbacks for sinks or basins must be properly installed, with sealed joints and edges to avoid the accumulation of dirt, grease etc.

Table-tops and other work areas, if no special surface is fitted, should be sanded and sealed, and the sealing treatment should be repeated periodically. Where specialized activity demands particular surfaces, or where other surfaces are fitted, care should be taken that joints and edges are sealed.

The construction and maintenance of kitchen areas and toilets are discussed separately below.

8.2.3 Kitchens and food safety

The health and hygiene of kitchens and food handling are subject to the requirements of the Health and Safety at Work etc. Act 1974 and its Workplace (Health and Safety and Welfare) Regulations 1992, and the Food Safety Act 1990 and its Regulations, the Food Hygiene Regulations 1991. Registration under the Children Act 1989 will involve an assessment of standards of areas for food preparation.

The Health and Safety at Work etc Act and its regulations (see above) impose a duty on employers for the general cleanliness of premises, and for the maintenance of suitable hygiene standards for facilities where employees prepare or eat their own food.

The Food Safety Act 1990 and its Regulations set standards for the safe handling and preparation of food. These are listed in Chapter 4. and include consideration of standards of equipment, training of staff, storage, preparation and cooking and general cleanliness.

- Equipment – must be appropriate for the job required and in good condition.
- Storage – the safe storage of food requires refrigeration below 5 °C. Frozen foods must be stored at minus 10 °C. Cooked and raw foods must be stored separately.
- Cooking – a cooking temperature of 63 °C is required, and up to 70 °C for a joint of meat. Eggs must be cooked until the yolk and white are solid. Food must not be reheated more than once.
- Cleanliness – all surfaces are to be kept clean, using a germ-killing product daily. Use separate equipment for each type of food prepared, or wash in between. Staff and children who prepare food must wash in warm soapy water and wear clean protective clothing. All equipment must be cleaned immediately after use and stored properly.
- Training – for food handlers is widely available, and is a requirement for staff working at sites where the preparation of food is part of the regular programme.

For adventure playgrounds, good practice in food preparations should include:

- all those involved washing their hands under running water before handling food, and drying them on individual towels or under hot-air dryers;

- no one with an infectious/contagious illness or skin disease handling food;
- all cuts and sores covered with a waterproof dressing;
- the wearing of clean clothing;
- the preparation of cooked and raw food in separate areas;
- washing fresh fruit and vegetables;
- thoroughly defrosting frozen food;
- keeping food covered and either refrigerated or piping hot;
- washing up must be done thoroughly in hot water with detergent and wearing rubber gloves, it must then be air dried;
- the supervision of children at all times.

The preparation of food at camps or barbecues etc. should follow these standards as far as is possible and reasonable.

All cooking and kitchen utensils must be kept clean and not be used for other purposes. Washing up equipment and tea towels must be replaced regularly. Any foodstuffs kept in the kitchen must be stored in clean, vermin-proof receptacles. Waste foodstuffs must be disposed of effectively.

The kitchen must be designed and fitted so as to facilitate cleaning and avoid contamination of all surfaces which should be of impervious materials, such as stainless steel or plastic laminate, to facilitate cleaning and avoid contamination. Cookers and other equipment should be moveable, not built in, to enable cleaning behind them.

More details on food hygiene can be obtained from the local authority environmental health department.

8.2.4 Toilets

The provision and condition of sanitary conveniences and washing facilities are subject to the requirements of the Health and Safety at Work etc. Act 1974 and its Workplace (Health and Safety and Welfare) Regulations 1992. Registration under the Children Act 1989 will involve an assessment of the standard of these facilities.

The Health and Safety Regulations (above) are concerned with the employees' duty to provide suitable and sufficient sanitary conveniences and washing facilities for employees (see Chapter 4).

Children Act Guidance and Regulation is principally concerned with the welfare of children and requires that toilet and washing facilities are provided at premises where children of 5–7 years are catered for. A ratio of one toilet and washbasin for ten children is recommended for open access schemes, with separate facilities for staff. The provision of separate facilities for boys and girls and staff is normally acceptable for adventure playgrounds. Where large numbers of children attend, it is not necessary to increase the number pro rata. The aim should be to ensure that young children have access to a toilet when they need it and to a washbasin nearby.

All toilets must be cleaned at least daily. Lavatory bowls and urinals must be brushed down and treated with a toilet cleaner or bleach. Seats must be cleaned with disinfectant. Basins must be cleaned and taps wiped with disinfectant. Walls and floors must be washed down with disinfectant regularly and their surfaces must be well maintained. Particular attention must be paid to the area surrounding the waste pipe outlet from the pan as this accumulates dirt and is often not easily accessible to a mop or broom.

There must be an adequate supply of toilet paper. Facilities for the disposal of sanitary towels must be available.

Washbasins must have hot and cold running water and soap available. The temperature of the water must be properly regulated and not exceed 30 °C. Paper towels or electric warm air hand driers are recommended. Roller towels are not recommended as they have been known to provide a hazard to younger children.

Children with certain disabilities may need to be regularly toileted, washed and changed. There should be facilities to do this in comfort. It is important therefore to keep a supply of clean clothes for use in emergencies.

8.2.5 Heating, ventilation and lighting

The temperature, ventilation and lighting of an indoor workplace is subject to the requirements of the Health and Safety at Work etc. Act 1974 and its Workplace (Health and Safety and Welfare) Regulations 1992. Registration under the Children Act 1989 will involve an assessment of heating requirements, ventilation and lighting.

The Health and Safety Regulations (above), are concerned with the employer's duty to provide heating levels which give reasonable comfort without the need for special clothing (see Chapter 4). This is normally at least 16 °C. However for an adventure playground building a temperature of 20 °C is more appropriate. One hour after opening, the temperature inside the building should not be less than 15.5 °C. Where children under five years of age are involved, a higher temperature is required.

Children Act Guidance and Regulation gives no specific guidance on the heating, ventilation or lighting requirements of open access facilities such as adventure playgrounds. However in assessing the fitness of buildings as part of the registration process, the local authority will consider these as part of the assessment of the suitability of premises. Actual requirements will be determined locally but are most likely to follow the information contained in the Health and Safety Regulations outlined above.

Heating appliances must be kept in proper working order and suitable protected to keep children safe. They must be regularly serviced by a qualified and competent person. Portable heaters such as paraffin or calor-gas stoves are not recommended. There should be adequate circulation of fresh air throughout the building, especially in kitchens and toilets.

8.3. THE OUTDOOR SITE

8.3.1 Children's constructions

Children's camps and dens must be checked daily for hygiene. The nature of children's impulses to build and their sometimes limited technical ability mean that they may use unsuitable materials or construction techniques.

Children should be encouraged to:

- provide adequate ventilation in their camps and dens; the use of plastic or polythene sheeting for waterproofing can restrict ventilation, as well as creating a real fire hazard;
- avoid the use of unsuitable materials, particularly for furnishing; these include mattresses and upholstered chairs; these will be a fire hazard and may be, or become, verminous or mouldy.

8.3.2 Cleaning

It is a requirement of the Workplace (Health Safety and Welfare) Regulations 1992 that floor surfaces and traffic routes must be of sound construction and free from obstructions, holes, slopes and uneven slippery surfaces which are likely to cause a slip, trip or fall.

The site should be clear of rubbish of all sorts. If refuse is allowed to accumulate it can attract vermin and represent both health and fire hazards. All playground users should be encouraged to deposit litter, particularly food wrappers of left-overs, in rubbish containers which should be adequate in number, clearly marked and emptied regularly. The containers must not be allowed to overflow.

A central container for non-combustible rubbish should be provided. A skip, or dustbin are the most appropriate containers which must, where possible, have lockable covers. Refuse should be removed weekly but if this is not possible or necessary, containers should be disinfected weekly with dustbin sanitizing powder or liquid disinfectant.

Where used clinical needles (sharps) are likely to be found as rubbish they must be handled with extreme care and placed into approved sharps boxes. More details on safe practice should be obtained from clinical health-care workers locally.

Regular cleaning and maintenance will be required, as set out in the section on buildings above. Attention to risks to employees as a result of manual handling may be required if cleaning and maintenance involves lifting and carrying of heavy items such as timber, pallets etc.

Cleaning staff must be properly trained and supervised.

8.3.3 Animals

Animals are a year-round commitment and require specialist care based on expert knowledge.

They have the following basic needs which must be met at all times:

* freedom from thirst, hunger and malnutrition;
* suitable comfort and shelter;
* the prevention of, and rapid treatment of, injury, disease, infection or infestation;
* freedom from fear;
* freedom to display normal behaviour.

Where animals are kept on site, they must be housed in shelters of adequate size and appropriate design for the species. Such shelters must be well-ventilated, well lit and provide a good environment for their occupants. Where appropriate, animals must be given access to the natural environment, in protected runs or enclosures with access to sunlight, fresh air, water, grass and earth.

Animal shelters must be cleaned thoroughly and regularly. Excreta and soiled bedding or nesting material must be removed and disposed of effectively (see section on waste disposal below). Where this material is used for compost, compost heaps must be sited away from activities and properly contained.

The playground's animals must regularly be inspected and treated for worms, fleas and other parasites. The playground's animals must be checked by a vet or other qualified person at least twice a year.

Other animals (particularly dogs) must not be allowed on site. There is an unacceptable risk of children being attacked or the playground being fouled. Excreta, as well as being an unpleasant hazard to children at play, may contain organisms dangerous to health.

Children and other visitors coming into contact with animals on site should be asked to:

* respect the animals – not to chase or frighten them;
* be safe – not to enter any paddock or animal enclosure without express permission, close and lock gates;
* be careful – because the behaviour of animals cannot be predicted, especially where small children are concerned;
* be hygienic – by washing hands after touching animals;
* not feed animals or drop litter, as animals may try to eat it.

A warning should be given to women who are, or may be, pregnant to avoid contact with sheep and lambs at lambing time.

8.3.4 Pesticides

Where pesticides are stored or used on site, the Food and Environmental Protection Act 1989 Part III – The Control of Pesticides Regulations will apply.

These Regulations are concerned with the storage and use of pesticides. Pesticides are substances used for protecting plants and killing bugs etc. They may appear on adventure playgrounds as weedkiller or timber preservative.

As pesticides are also likely to be identified as hazardous substances in a general risk assessment they would also be subject to assessment under the COSHH regulations.

All reasonable precautions must be taken to protect the health of human beings, creatures and plants, to safeguard the environment, and in particular to avoid pollution of water.

If such substances are used on an adventure playground then specific advice and guidance must be sought on their storage and use.

8.3.5 Drainage

Where drainage inlets to sewers, or other drainage (i.e. from sinks or rainwater pipes) exist or are installed, they should be securely protected from damage. They must be covered with a suitable grille or mesh to prevent blockage and, where appropriate, treated periodically with disinfectant, and flushed through once or twice a year with a jet from a hose pipe.

Manhole or inspection covers for main drains must be solid and secure. Covers and their concrete surrounds should be regularly inspected and maintained. They must not create a hazard by projecting above pathways.

8.4 DISPOSAL OF WASTE MATERIALS

It is a requirement of the Workplace (Health Safety and Welfare) Regulations 1992 that waste material is disposed of properly. There is also a duty of care to keep waste safe.

The disposal of non-domestic or hazardous rubbish is subject to the requirements of the Environmental Protection Act, 1990 Part II. It must only be disposed of to authorized persons such as licensed carriers. Details of the types of waste involved must be given to the carrier. The burning of rubbish is not permitted.

Such waste can be anything which an adventure playground wishes to dispose of. Such waste should be:

* stored securely;
* only passed on to someone who has the authority to take it, such as council waste collectors, registered waste carriers;
* accompanied by a written description of the waste (except household waste).

Refuse from food preparation should be put in containers with lids in place at all times.

Waste-collection authorities are required, under Section 49 of the Environmental Protection Act 1990 Part II, to draw up and execute plans for the re-cycling of waste. Adventure playgrounds should use opportunities for recycling whenever they are appropriate.

More information on the disposal of waste is given in *Waste Management, The Duty of Care, A Code of Practice*, HMSO, 1996.

8.5 HEALTH

Playworkers must at all times seek to protect the general health of playground users. They must actively encourage children in hygienic practices (for example, washing hands after using the lavatory or handling the animals) and, where necessary, develop awareness of remedies to problems of personal hygiene. They must be able to recognize the symptoms of common infectious or contagious diseases or conditions and take action to prevent their spread through contact at the playground. Of particular relevance are the following:

> lice and fleas, mumps, measles, scabies, german measles, eczema, chicken pox, impetigo, whooping cough, ring and tape worm, conjunctivitis, jaundice, meningitis, scarlet fever. (See Appendix 9 for exclusion periods for common infections.)

Playworkers should be aware of the serious implications for health of drug abuse, and conscious of the 'symptoms' of such abuse. Under no circumstances must the use of drugs, including alcohol, be permitted on the playground. Playworkers should be aware of the possible courses of action open to them in dealing with children or other playground users who are using, or are under the influence of, drugs. Of particular concern should be alcohol, solvents, amphetamines, barbiturates and opiates, all of which cause serious physiological damage as well as behavioral modification.

Under no circumstances must any child or adult ever be allowed to attend the playground while under the influence of alcohol or any controlled or non-prescribed drug. Judgement and physical co-ordination may be impaired and the likelihood of accident is increased.

Tobacco is also a drug, damaging to health and smoking is a potential a fire hazard. It is illegal for children under 16 years old to buy tobacco. It should be forbidden on an adventure playground, and that prohibition should include any adults on the site.

There is a section in Chapter 9 on child protection.

8.6 EQUIPMENT, DEVICES AND SYSTEMS

The Health and Safety at Work etc. Act 1974 and its Workplace (Health Safety and Welfare) Regulations 1992, requires that any equipment, devices and systems which are used in order to comply with those regulations must be maintained in efficient working order and good repair.

In the context of adventure playgrounds this will include, for example, heating, lighting and plumbing systems, cleaning equipment and emergency lighting.

8.7 LEGISLATIVE BACKGROUND

Part Two of this publication contains summaries of the following legislation referred to in this chapter:
- The Health and Safety at Work Act (1974), and its Regulations:

  - Control of Pesticides Regulations 1986
  - Control of Substances Hazardous to Health (COSHH) Regulations 1994
  - Workplace (Health and Safety) Regulations 1992
  - Manual Handling Operations Regulations 1992
  - Personal Protective Equipment at Work Regulations 1992
  - Management of Health and Safety at Work Regulations 1992

- Food Safety Act 1990
- Food and Environmental Protection Act, 1989 Part III – The Control of Pesticides Regulations 1986
- Environmental Protection Act 1990 Part II

8.8 FURTHER READING

The following are the main source documents for this chapter:

Association of City Farms, *Policy on Animal Care*, undated.

Food Sense (1990) *The Food Safety Act 1990 and You – A Guide for Caterers and Their Employees*, Food Sense.

Food Sense (1992) *The Food Safety Act 1990 and You – A Guide for the Food Industry*, Food Sense.

Guidelines for Good Practice for Sessional Playgroups, Pre-School Playgroups Association 1989.

Health and Safety Executive information leaflet (1970) T*he Food Hygiene (General) Regulations*, HMSO.

Health and Safety Executive (1993) *Step by Step Guide to COSHH Assessment*, HMSO.

Health and Safety Executive (1993) *COSHH, A Brief Guide for Employers*.

Kids' Clubs Network *Food Safety and Kids' Clubs*, undated.

Ministry of Agriculture and Fisheries *Best Before and Use by – A Guide to the Changes*, undated.

The Children Act Guidance and Regulations, *Volume II – Family Support, Day Care and Education Provision for Young Children*, HMSO, 1991.

Waste Management, The Duty of Care, A Code of Practice, HMSO, 1996.

Chapter 9

Accidents and emergencies

9.1 INTRODUCTION

All playgrounds must be prepared for, and able to deal with, possible emergencies. This will require appropriate equipment, and agreed procedures. The most likely emergencies will have been identified as part of the general risk assessment undertaken in accordance with the Management of Health and Safety at Work Regulations 1992. Registration under the Children Act will include an assessment of emergency procedures.

9.2 PROCEDURES

Agreed procedures for dealing with the most likely occurrences must be established and adhered to. Such procedures must be communicated to all staff, volunteers and playground users. All staff and others likely to deal with incidents must receive adequate training and instruction in the necessary skills, the use of equipment and all relevant procedures.

Accidents and emergencies occurring on site will almost always require additional specialist help. Every playground must have a telephone installed on site. Where the telephone is locked, it is essential that there is ready access to it during hours of opening and that emergency calls, including 999 calls, can be made at any time.

All workers must be aware of the procedures for emergency calls. They must be able to identify the address and location of the playground clearly, and must be able at all times to give as many details of the emergency as possible. It is advisable to give the local police, ambulance and fire service an accurate map of how to reach the playground and the precise location of the access points.

All access gates must be kept clear at all times, both inside and outside the playground. A designated person should meet the emergency services upon arrival.

Dial 999 first for incidents requiring police, fire or ambulance services. Addresses and telephone numbers of local police stations, fire station, hospital accident and emergency departments and doctors' surgeries, gas, electricity and water companies should be clearly displayed by the telephone (and kept on a card, in or by, the first-aid box).

9.2.1 Child protection

One emergency situation that adventure playgrounds may face is the sudden discovery of child abuse. It is important that effective and known procedures are in place, so that the child is not exposed to further risk and that the situation is dealt with calmly and fairly. There is much useful material and information available on this subject and readers are encouraged to access these via local sources, particularly the Social Services Department Child Protection Team of their local authority. The Department of Health guidance document, *Working Together* (available from HMSO) is useful for voluntary-run projects.

Playworkers are in a prime position to notice changes in behaviour, or physical signs, or to obtain verbal information, which may give cause for concern for the safety and welfare of children. It is therefore imperative that adventure playgrounds have a policy about what to do in the event of disclosure by a child or suspicion of abuse. They must know their Social Services contact if required.

In the first instance the child must always be believed and her/his information acted upon. A child must be supported, but be made aware that such matters cannot be kept secret. There must be a procedure to deal with suspected non-accidental injury, or sexual abuse. A trusted relationship with the local Child Protection Team is the key. Positive training by, or with, that team can help. All workers must be clear whom they report to when this information comes to light. Playworkers should not take it upon themselves to talk to parents.

This is a sensitive area and there should be plenty of expert advice and help available from local sources.

9.2.2 Fire

The Fire Precautions Act 1971, as amended by the Fire Safety and Safety of Places of Sport Act 1987 provides useful guidance on emergency procedures in respect of fire in buildings, and are summarised below.

Procedures

- All procedures should take account of the possible absence of personnel from premises, e.g. because of sickness, leave etc.
- The fire brigade should be called immediately, irrespective of the size of the fire.
- One person (but see above) should be given responsibility for ensuring that pre-planned action is carried out, and meeting the fire brigade upon arrival.

Equipment

- All persons should be trained to use fire extinguishers.
- Extinguishers suitable for the likely risks should be provided and be clearly marked. Types of fire extinguishers are listed in Chapter 4.

Alarm

- A manually operated fire alarm is required in licensed premises.
- On adventure playgrounds if a proprietary fire alarm is not fitted a suitable recognizable system to raise the alarm should be provided. This could be, for example, by whistle and command, or by bell and command. The fire service recommends a 12 cm bell.

These Regulations also make recommendations concerning means of escape and good housekeeping which are covered in Chapter 4.

Fire extinguishers must be readily accessible, regularly checked and maintained in accordance with manufacturers' instructions. Kitchens (and craft areas where appropriate) should be especially considered. It is advisable to provide a fire blanket in these areas, as well as fire extinguishers of the dry-powder or other variety suitable for use on electrical or fat fires.

Appropriate smoke alarms must be fitted in passageways and strategically placed throughout the building. A heat-sensitive alarm is recommended in the kitchen, where smoke alarms are not practical. The advice of a fire officer must be sought on the location of these. Smoke alarms must be checked and maintained as follows:

- once a month – check the battery by pressing the test button and test the sensor by holding a recently extinguished taper or candle under the alarm;
- once a year – change the battery and vacuum the dust from inside the alarm.

Inside buildings, attention must be paid to exits which must be unobstructed and clearly marked. External doors must not be locked while the building is in use and 'crash-bars' or an alternative agreed with the fire officer must be fitted if certain doors are to be kept closed. Auxiliary lighting may be required. The advice of the local Fire Officer must be sought before buildings are used by children.

All staff must be instructed in the use of all fire-fighting equipment. They must be made aware of which items of equipment to use for the various types of fire. They must be aware of any established routines for dealing with fires and details of the fire code must be displayed prominently in the building.

In the event of a fire in the building the alarm procedure must be carried out. The fire brigade must be called no matter how small the fire. The building and/or on the site must be evacuated. Each area of the building and/or site must be checked by the designated person according to the agreed procedure, to ensure that all users are clear, so long as this checking will not endanger the person doing it. If it is safe to do so the records of users must be removed. An assembly point will already have been chosen. It must be at least 25 m from the building. All people must meet there. Playworkers must assess whether the clearing of the building has been undertaken completely, through consultation between themselves and by checking with those who have been evacuated. A roll-call may be possible. The designated person must meet the fire brigade on arrival.

Arrangements for notifying parents and releasing children from the site must be included in the agreed emergency procedure, and must take account of the possibility of children being distressed and confused by the experience.

It is recommended that fire drills be undertaken at least twice a year, and preferably every two months. The fire drill procedure must be clearly displayed and all members of staff must be given a copy. Fire drills must be recorded.

9.2.3 Mains services

All staff must know the location of mains gas and water stopcocks and electricity switches. These must be readily accessible to staff at all times though secured against interference by children. Emergency contact numbers for gas, electricity and water companies must be displayed by the telephone, or in another prominent position.

All mains service installations should be regularly inspected and serviced by qualified professionals in accordance with manufacturers' instructions and the requirements of the relevant authority.

All modifications or repairs to any mains service equipment or installation must be carried out only by competent craftsmen. Improvised repairs must never be carried out. Until repairs are carried out, any damaged or unserviceable appliance or part of an installation must be disconnected or, if this is not possible, and any hazard exists, the whole system must be isolated from the mains supply.

All work on mains installations or systems must be carried out only when the part of the system affected is isolated from the mains supply.

9.2.4 First-aid

Inevitably there will be accidents to children using the adventure playground. Most of them will result in minor bumps, scrapes or cuts only, but the workers must be prepared to deal with more serious injuries.

The Health and Safety at Work etc. Act, through its First-Aid regulations, requires employers to make adequate provision in the event of accident to and/or illness of their employees, and for others who may come into contact with their work. The Children Act 1989 Guidance and Regulation, Volume II requires at least one person working in a day care-setting to have received first-aid training.

There must be a qualified first-aider on site, and all full-time staff must have a current first-aid qualification from a recognized body. Certificates are valid for only three years and a refresher course must be taken to maintain the validity of the qualification. The qualification must be gained through completing the full course.

Where training does not include attention to first-aid for children, including resuscitation of children, these must be provided as additions.

Only a parent, or someone having long-term care of a child, can give consent to non first-aid medical treatment, i.e. which a doctor has identified as being required. This consent cannot in any way be given to those working with children. Every playground must be properly equipped to deal with injuries. A separate area of the building should be set aside and furnished for treatment, see Chapter 4.

The playground's first-aid equipment must be kept in a secure container clearly marked with a white cross on a green background. The first-aid box must be:

- made of suitable material and designed to protect the contents as far as possible from damp and dust;
- kept in an accessible place, but secure from interference by children;
- always replaced after use;
- always kept tidy and clean;
- regularly replenished – a stock list should be kept inside the door or lid to enable a check on contents.

It should be borne in mind that certain items may deteriorate with age. All items must be checked regularly and replaced where necessary.

If the first-aid box is not portable, consideration should be given to the provision of a pack of essentials which could be taken out of a first-aid area onto the site (if for example, an injured child cannot be moved while awaiting medical attention).

There should be a suitable refuse container which is frequently emptied. Used dressings, wipes etc. should be disposed of in a sealed plastic bag. There must be at least one blanket which is stored away from damp or dust.

9.2.5 First-aid equipment

The Health and Safety Executive recommends contents for first-aid boxes, and these are listed at Appendix 10. This includes a bottle of sterile water.

It is good practice to take a travelling first-aid kit for off-site activities. The recommended contents of a small first-aid kit are given at Appendix 11.

Where tap water is not available, sterile water or sterile normal saline, in disposable containers each holding at least 300 ml, must be kept easily accessible and near to the first-aid box, for eye irrigation.

9.2.6 Hygiene and protection from contamination

In order to prevent possible contamination from blood or body fluids (see below), the following items must be provided:

- sterile seamless latex or vinyl gloves;
- disposable plastic aprons;
- eye protection;

- disposable towels;
- yellow clinical/biohazard sacks;
- Presept granules and tablets;
- soap and water.

9.2.7 Blood and body fluids

Workers who come into contact with blood and/or body fluids may be exposed to occupational risk from blood-borne viral infections such as HIV or Hepatitis B. The most likely means of transmission of these viruses is through accidental inoculation of infected blood by sharps (clinical needles), or by blood or body fluids splashing onto broken skin or mucous membrane. Since it is impossible to identify all those who are seropositive to HIV or Hepatitis B, as a matter of good practice playworkers must use barrier methods when giving first-aid.

The following advice for playworkers is based on DoH (1990), *Guidance for Clinical Health Care Workers, Protection Against Infection with HIV and Hepatitis Viruses: Recommendations of the Expert Advisory Group on AIDS,* London, HMSO.

1. Cuts or abrasions on the person of the first-aider must be covered with a dressing that is waterproof.
2. Sterile seamless latex or vinyl gloves must be worn where there may be contamination of hands by blood or body fluids, and discarded after first use.
3. Hands must be washed in warm soapy water after the treatment is concluded, even if gloves are worn.
4. Disposable aprons/eye protection must be worn where there is a possibility of splashing by blood or body fluids.
5. Spillages of blood or body fluids must be dealt with by a person wearing gloves and apron (as above), the spillage must be covered with disposable towels to soak up excess, wiped up and the surface treated with the appropriate domestic cleaning product for that surface. Full details of this procedure are provided in Appendix 12, and include the use of Presept granules and tablets.
6. All contaminated waste must be placed into yellow clinical sacks and arrangements made for their incineration (the Environmental Health Department of the local authority will have information on disposal arrangements).

9.2.8 Notifying parents/guardians

Parents/guardians must be notified of any first-aid treatment given. How this is undertaken will depend upon the type of injury, but may be in person, by telephone or as a note to be taken home by the injured child. Any injury/treatment concerning the head must be notified in person or on the telephone.

In the event of any child requiring immediate further treatment, whether by their own or a local GP or at the Accident and Emergency department of a hospital, their parent or guardian must be contacted and asked to accompany the child. If this is not possible, a playworker must accompany the child and arrangements to close the site be made if this affects satisfactory staffing levels. Strenuous efforts must be made to contact the parents or guardians.

Only a parent or someone having long-term care of a child can give consent to medical treatment which a doctor has identified as being required, and this consent cannot in any way be given by those working with children.

Full details of the accident or injury and any first-aid treatment given must accompany children being moved off site for any further treatment.

Where further treatment is required, but is not urgent, clear details must be given to parent(s) or guardian(s).

9.2.9 Reporting accidents

Every incident leading to injury, and 'near misses', must be recorded. Reports of accidents must be made at the time of treatment. Where, as sometimes happens, a child leaves the site before treatment is possible, a note of the incident must be taken and workers must follow up to ascertain the extent of injury and treatment received. A suitable means of recording incidents must be kept in an accessible place, preferably with the first-aid equipment. Reports must be made in triplicate, with a copy held at the playground, and copies available to parent(s) or guardian(s), and to medical practitioners responsible for further treatment.

The report must include:

- name, address and telephone number of the adventure playground;
- name, age and address of injured person;
- date and time of incident;
- nature of injury;
- details of treatment given/action taken, and person treating;
- details of incident, including location, cause (where appropriate) and name/s of witness/es;
- if the injured person is to be treated elsewhere, the report must give details of any special conditions, allergies or handicaps or medication being taken, where known;
- where the injured person is not to be treated elsewhere immediately, the report must contain for the information of parent(s)/ guardian(s) any recommendation on further action (e.g. observation, examination by doctor, tetanus immunisation, advice on care of dressing).

A sample accident report form is included as Appendix 13.

The Reporting of Injuries, Diseases and Dangerous Occurrences Regulations (RIDDOR) (1985), places three principle duties on employers:

1. the keeping of records of accidents at work – an accident book;
2. the reporting of accidents at work, including to those who are not employed at the place of work;

3. the reporting to the Health and Safety Executive of dangerous occurrences (near accidents).

Reporting is to be made to the appropriate enforcing authority for the premises. For adventure playgrounds operated by local authorities, this is to the Health and Safety Executive, for others it is to the Environmental Health Officer.

The reporting of accidents is related to the extent and nature of the injury suffered. Reportable accidents relevant to adventure playgrounds include:

- fractures of the skull, spine or pelvis;
- fractures of the arm, wrist, leg or ankle;
- amputation of the hand, foot, finger, thumb, toe or any part thereof;
- loss of sight of eye, penetration injury to the eye, chemical or hot-metal burn to eye;
- injury (including burns) requiring immediate medical treatment, or loss of consciousness because of electrical shock;
- loss of consciousness due to lack of oxygen;
- acute illness/loss of consciousness resulting from absorbtion of any substance by inhalation, ingestion or through the skin;
- any other injury which results in admission to hospital for more than 24 hours;
- for employees only, absences due to injury at work where these absences are for more than three days;
- fatalities – the reporting to the Health and Safety Executive of deaths within one year as a result of a reportable injury. All fatal accidents, major injuries and dangerous occurrences have to be reported immediately (e.g. by telephone).

Records of children regularly using the playground must include:
- name/address/telephone number/date of birth of child;
- at least one emergency contact number;
- any recurrent health problems, e.g. allergies/asthma/epilepsy, medication (see below for discussion on administration of medicines);
- if possible, include the name and telephone number of the general practitioner;
- dietary requirements and food allergies;
- school/headteacher/teacher/social worker/other agency;
- position in family.

Any records kept on computer are subject to the Requirements of the Data Protection Act 1984.

9.2.10 Medication

Children who require the administration of prescribed drugs at times when they attend an adventure playground would be unfairly discriminated against if staff refused to do so, and as a consequence the children would be prevented from attending.

In addition the administration of medication can be life-saving, for example anti-asthma treatment. However the administration of drugs must not be undertaken unless a proper procedure has been devised. Such a procedure would be drawn up and agreed in consultation with the medical services and would require:

- consent from parents and GP, which would need to be very clear and specific;
- the child to be responsible for administering their own medication; or,
- if not capable of doing so the administration to be supervised by a nominated key worker for the child;
- a record of each administration.

Until such a procedure is in place, no drugs, even junior aspirin, may be dispensed by playworkers.

9.3. VIOLENCE TO STAFF

Violence can be described as any incident in which a person is abused, threatened, or assaulted, irrespective of whether or not physical injury is sustained. Adventure playground staff my find violence directed at themselves.

The law only allows for employees to put up such self-defence as is reasonable to protect themselves, given the circumstances.

Employers are responsible for the health, safety and welfare of staff and therefore have a duty to ensure that safe systems of work are in place which ensure that employees are not put at unnecessary risk, and that staff are adequately trained to recognize, and if possible deal, with confrontations and violence.

Employers must ensure that:

- all employees are trained to deal with situations that they may encounter during their normal work;
- that all potential 'at-risk' situations are assessed from a personal safety viewpoint, and appropriate measures implemented as required;
- there is an agreed reporting procedure for incidents involving violence, and that employees are familiar with it.

Employees must ensure that:

- they advise their supervisor/manager if they are concerned about their personal safety;
- they take reasonable care for their own personal safety and co-operate with their employers in the operation of safe systems of work, including undertaking training;
- they co-operate with their employers in the investigation of any serious incident reported by them.

Adventure playground managers should undertake a risk assessment concerning violence to staff. This must consider the work environment, work methods, the nature of the work to be done and the methods by which it is carried out.

Home visiting may be a regular part of the work of a playworker. This can leave the worker more vulnerable, and the following safeguards are useful:

* use knowledge of the families to be visited to assess how/whether to undertake a home visit;
* information must be provided to colleagues concerning when, where and by whom the visit is to be undertaken;
* carry a means of identification;
* do not enter if the person they wish to see is not there.

Night call-outs to sites may be required of keyholders. The keyholder should establish that the police have been called to the site and will be there when they arrive. Provision must be made for safe transportation to and from the site. The police may help with this.

An incident reporting procedure must be in place which identifies all violence or threats of violence to staff. An assessment of all reported incidents should be made in order to improve strategies for prevention.

Training must involve dealing with the public, specific procedures for, e.g. home visiting, detecting warning signs, and understanding how their own behaviour, attitude and appearance can affect a situation, and advice and instruction on the use of alarms etc.

9.4 LEGISLATIVE BACKGROUND

Part Two of this publication contains summaries of the following legislation referred to in this chapter:

* The Health and Safety at Work Act (1974) and its Regulations;
  * Health and Safety (First-Aid) Regulations 1981;
  * The Reporting of Injuries, Diseases and Dangerous Occurrences Regulations (RIDDOR) 1985;
* The Management of Health and Safety at Work Regulations 1992;
* Fire Precautions Act 1971, amended by the Fire Safety and Safety of Places of Sport Act 1987;
* The Children Act 1989.

9.5 FURTHER READING

The following are the main source documents for this chapter:

A Fire Survival Guide – What To Do if Fire Breaks Out in Your Home Home Office, Fire Safety in the Home leaflet, undated.

Health and Safety Commission (1991) *First Aid at Work – Health and Safety (First-Aid)*.

Health and Safety Executive (1988) *Preventing Violence to Staff.*

Health and Safety Executive (1992) *Management of Health and Safety at Work Approved Code of Practice.*

Protect Your Home From Fire (1994) Home Office Fire Safety in the Home leaflet.

Reporting of Injuries, Diseases and Dangerous Occurrences Regulations (RIDDOR) (1985).

Royal College of Nursing (1993) *Universal Precautions*.

Southampton City Health and Safety Information Bulletin Number 4 (1981) *The Health and Safety (First Aid) Regulations 1981*.

Southampton City Council (1982) *Safe Working Procedure No 28 – The Prevention of Violence at Work*.

Wake up! Check Your Smoke Alarm (1992) Home Office information leaflet.

Chapter 10

Activities off the site

10.1 INTRODUCTION

Once children are taken off the site as part of organized activities, they are exposed to a range of risks over which playworkers have limited control. In order to ensure that all reasonable steps are taken to eliminate or reduce the impact of these risks, principles of safe practice must be applied.

This chapter outlines these principles. The information set out here should be treated only as a guide, and specialist advice must be sought where appropriate, particularly with regard to inherently hazardous activities.

Activities organized away from adventure playgrounds are subject to the requirements set out in the Health and Safety at Work etc. Act 1974 and the Management of Health and Safety at Work Regulations 1992.

The Health and Safety at Work etc. Act places general duties on all people at work to carry out their work in a way which ensures, as far as is reasonably practicable, that persons affected by it are not exposed to risks to their health and safety.

The Management of Health and Safety at Work Regulations 1992 require employers to make a suitable and sufficient assessment of the risks arising from their undertaking. Details of risk assessment are given in Chapter 2.

The organisation and operation of off-site activities must be subject to risk assessment as required by the Management of Health and Safety Regulations. This assessment must identify the potential risks to which participants will be exposed and identify what steps are required in order to eliminate or reduce those risks.

The Children Act 1989, through its Regulation and Guidance Volume II, also makes recommendations concerning the organization and provision of off-site activities.

10.2 IDENTIFYING RISKS AND PROMOTING SAFETY

The key to risk assessment is the collection of information and the application of knowledge and experience. In the case of off-site activities, playworkers will frequently need to seek and collect information and knowledge from outside sources and learn from, or rely on, the experience of others in order to undertake a risk assessment.

A broad range of information will need to be collected and considered concerning hazards which may be associated with the activity. These may include information:

- on routes – from police, coastguard, motoring organizations;
- on places to be visited – from operators of the facilities to be visited, designated persons within the organization, previous users, advisory bodies, published advice, policies, personal visits;
- on activities to be undertaken – from written policies or designated persons within the organization, governing bodies, personal visits, personal experience, previous users, activity organizers.

Information may be collected by:

- written enquiries, telephone enquiries, personal visits/interviews, interviews with previous participants including children.

The information collected must be recorded for future information and as a record of the collection of the information.

Once collected, it should be considered in the light of the needs and capabilities of the children who will be taking part in the activity. This will require an understanding of their development stages and behaviour and of their ability to assess and minimize risks for themselves, the type of activity, age, abilities, needs and culture of the children, the group size and the knowledge/experience of staff in respect of the specific activity to be undertaken.

10.3 PLANNING AND PREPARATION

Once a risk assessment has been completed, the activity will require planning and preparation.

A party leader must be chosen. This person will have overall control of the activity, its planning, preparation and execution. That person must be appropriately trained and experienced to lead the party. That does not mean that he/she must be trained and experienced in every activity to be undertaken by the party, as it is essential for any specialist and/or inherently hazardous activity to be lead by a competent expert in that activity.

The party leader will be responsible for reviewing the information collected and, if necessary updating it or obtaining further information. It is desirable that the party leader should have first hand experience of the places to be visited. This should preferably be by a recent prior visit. If this is not possible then a thorough assimilation of information obtained should be undertaken. In any event, the assumptions made by the party leader in advance of the activity should be reviewed by on-the-spot reconnaissance as the activity progresses.

A back-up plan must be devised which provides suitable alternatives. These must also be subjected to a risk-assessment process.

The party leader must have authority to exercise full responsibility throughout the planning, preparation and execution of the activity, even if, for example a person, with greater seniority within the organization takes part.

The party leader will be responsible for briefing helpers, children who will be taking part, and their parents.

Briefing is of utmost importance. For the helpers, it provides an opportunity to familiarize themselves with the potential risks and ways to eliminate or minimize them. The briefing should fully explain the responsibilities of the helpers and the chain of command. Good-housekeeping routines can also be explained.

The briefing of parents will fully describe the nature of the activities to be undertaken and the potential hazards identified. Information should be given on transport, staffing levels and competence and experience of any competent experts leading any inherently hazardous elements. The back-up plans must be described. Parents must be given information on the types and levels of insurance in effect for the children taking part. Details of departure and return times, emergency contact procedures and clothing/food etc. can be given at the briefing. More details concerning insurance arrangements are given below.

Not all briefings need be verbal. Written information will be adequate where the trip is to and/or involves activities of which the parent could reasonably be expected to have understanding and knowledge. Where provided it must contain all the information required for a verbal briefing (see above). A combination of verbal and written briefings may be appropriate in some circumstances.

Briefing the children will provide an opportunity to describe the trip and its activities, agree a framework of behaviour and identify where variations to the arrangements will more particularly meet the needs and interests of the children. Where leaders are not well known to the children taking part, the briefing must be arranged in a way which establishes useful contact and promotes goodwill.

Briefing must continue throughout the off-site activity. This will provide opportunities for safety rules to be reinforced and give opportunities for children to contribute their views and opinions. The party leader will be able to use these briefings to give information about any changes to the procedures which he/she has identified as part of the on-the-spot reconnaissance (see above).

All briefings represent an informal exchange of information between the party leader and the children taking part and their parents/guardians. There must, in addition, be a formal recorded exchange of information.

Although this formal exchange is commonly called 'the consent procedure', where parents/guardians are asked to give consent in respect of an activity, it is more properly an exchange of information between the parties.

The party leader must provide parents/guardians with written material containing the information given during the verbal briefing. In return, the parent is asked to give information to the party leader which will assist him/her in ensuring, as far as is reasonably practicable the health and welfare of the child concerned. This may be information, for example, on a medical condition which may affect the child whilst undertaking the activity, or about water confidence/swimming ability. The parent/guardian will also be asked to consent to the child taking part in the off-site activity as it has been described in the information given. A model consent form is set out at Appendix 14.

The question of consent to medical treatment is discussed below.

The preparations will require the establishment of an emergency contact procedure. This must be two-way, allowing for contact to be made from the party leader to the parents and from the parents to the party leader. Both exist to be used in emergency situations.

This procedure requires an accurate list of all those taking part in the activity to be held and accessible throughout the duration of the trip both on the trip and back at base.

A nominated person or persons at base will hold in addition to the list of participants, details of how to contact parents/guardians of the children and helpers on the trip. That person will also have information which will enable them to contact the party leader at any time during the off-site activity. Where this activity lasts overnight or longer more than one person must hold this information.

The party leader, while away on the off-site activity, will have information on how to contact the nominated person or persons back at base for the duration of the trip, in addition to the list of all those taking part.

This consent procedure is relevant for all trips, no matter how routine. However, where trips take place regularly and routinely to a place where the risks are reasonably constant, e.g. a sports centre, parents/guardians may be asked to give consent to a series of visits.

10.4 HAZARDOUS ACTIVITIES

The procedure for preparation and planning outlined above is relevant for all off-site activities. However, some of these will involve activities which are inherently hazardous. These will be highlighted as a result of the collection of information and risk assessment and include water-based activities (for example, swimming, sailing, canoeing, rowing, hiking, hill walking, rock and mountain climbing, pot-holing and caving).

Once any hazardous activities have been identified, the party leader is required to ascertain the appropriate levels of experience and training required to lead the activity concerned, and, where the activities are to be operated from an activity centre, the suitability and safety of these facilities.

10.4.1 Expert leadership

Many local authorities have specialist staff, often located in Education Departments/Youth Service, who have expertise, and operate safety systems, which enable them to identify appropriate levels of expertise and give advice on the availability of expert leadership for hazardous activities. Where this is the case, codes of practice will usually be in operation. These may be binding on adventure playgrounds if they are local authority based, but in any event will provide useful expert advice and guidance.

For adventure playgrounds operated outside a local authority, or ones where expert staff are not available to advise on hazardous pursuits alternative sources of information and advice will be required. These include governing bodies of the activities to be undertaken and Duke of Edinburgh Award Scheme officers, who are usually county-based.

It is also good practice to find out if expertise from within a local authority can be used as a resource for advice and information, and to obtain copies of their codes of practice.

Once the appropriate level of expertise for the activity concerned has been established, it is good practice to view the certificates of attainment of any expert leader whose services may be employed.

10.4.2 Activity centres

Some hazardous activities organized from adventure playgrounds are operated from, and take place at, outdoor activity centres.

Any outdoor activity centres which are run in return for payment are to become regulated from October 1997. This regulation will take place under the Activity Centre (Young Persons' Safety) Act, 1995 as set out in the Adventure Activities Licensing Regulations, 1996. From that date adventure playgrounds should only use centres which have been registered under these Regulations.

The registration process will consider the number and qualification of instructors, the provision to the participants of information on safety, the provision and maintenance of proper equipment, and appropriate first-aid provision and medical back-up. Operators of activity centres will be required to undertake a proper risk assessment, and to have identified all the means required to protect, as far as is reasonably practicable, the safety of participants.

In the period up to October 1997, and where an activity centre is exempt from the registration scheme because it is not run in return for payment, adventure playgrounds considering using adventure activity centres must collect, as part of their risk assessment of the off-site activity proposed, information on the standards of operation of the centre to be used, by asking the centres:

- to state whether they have applied for registration; and if so,
- how their standards of operation meet the registration criteria; also,
- whether the qualification of their instructors/leaders matches those set out in the Regulations;
- the proposed ratio of instructors/leaders to participants;
- how they disseminate safety information to participants;
- what personal and other safety equipment is provided, and how it is maintained;
- whether the party will, at all times, have access to a person with a nationally recognized first-aid qualification.

More information on these elements is available from the HSE Publication, *Guidance to the Licensing Authority on the Adventure Activities Licensing Regulations 1996.*

As part of the preparation, the party leader must ensure that the liability insurance in respect of the adventure playground's activities provides cover for the hazardous activity to be undertaken. There is more information on insurances in the next section.

10.5 INSURANCES

Adventure playgrounds must consider all the risks arising from their activities and where the risks are great enough, to insure against them.

The only insurance which the operators of an adventure playground must by law take out is Employer's Liability Insurance. This is required by the Employer's Liability (Compulsory Insurance) Act 1969 which compels employers to insure themselves in respect of liability for injuries caused by their employees to their colleagues. This can and must be extended to include and cover volunteers. See Chapter 3.

In addition to Employers' Liability Insurance, it is strongly recommended that all adventure playgrounds must take out Public Liability Insurance. This provides indemnity against legal liability for injury to persons other than employees, and for damage to, or loss of, their property. It is important to give insurers details of the range and type of activities which are undertaken by the adventure playground and for which Public Liability Insurance is sought, including any off-site and/or hazardous activities. It is good practice to inform insurers in advance of any activity of a hazardous nature to be undertaken.

Some accidents, loss and damage occur which are no one's fault. Where no liability (fault) is involved, neither Employers' Liability insurance nor Public Liability insurance will provide compensation. If the operators of an adventure playground wish to provide insurance to cover no-fault situations, they can take out Personal Accident or Personal Accident and Sickness insurances. It is good practice to provide Personal Accident insurance for trips which extend overnight or longer, and for those who take part in hazardous activities.

10.6 SUPERVISION

The risk assessment will identify the supervision requirements of each particular off-site activity, taking into account all the potential risks. A party leader must be chosen to oversee the planning, preparation and execution of the activity (see above).

Off-site activities must never be supervised by one person alone and at least one of the supervisors must hold a recognized and current first-aid certificate. All supervisors must be fully briefed (see above).

The ratio of staff to children will have been identified as a result of the risk assessment and will vary according to the nature of the activity and the ages of the children, for example, activities outdoors will require closer supervision than an indoor visit.

In the case of day trips, the ratio of adults to children should normally be no less than 1:5 except in the case of children over 12 when the ratio should be 1:10 at the least. Any child under five years old must be accompanied by a parent or guardian who is not additionally responsible for the supervision of the remainder of the party.

In the case of residential trips, holidays or camps, a ratio of at least one adult to six children is required with a minimum of two adults. Mixed gender parties must be supervised by at least one male and one female.

10.7 PARENTAL CONSENT

No child may be taken on an outing unless a parent or guardian has returned a signed form of consent (see above). Consent must be obtained for each activity which will be carried out.

No consent should be sought, or can be given, in respect of the giving of permissions for medical treatment. Only a parent, or someone having long-term care of a child, can give consent to non first-aid medical treatment, i.e. which a doctor has identified as being required, and this consent cannot in any way be given to those working with children, through the wording of a parents consent form or otherwise.

The model consent form provided at Appendix 14 suggests appropriate wording.

Unless a procedure for the administration of medication has been devised and agreed, no drugs, not even junior aspirin, should be dispensed.

10.8 ADDITIONAL SAFETY PROCEDURES

The following additional good practice ideas will contribute to the safety of children taking part in off-site activities:

- children should be checked-off each time the coach or transport is boarded;
- each time the party arrives at a location, and each time the coach or transport is embarked from, the children should be told exactly when, and from where, the party will be leaving again;
- supervisors should each be made responsible for a definite group of children, should be issued with a written list, and children should be made aware of who is supervising the group they are in;
- the use of name badges should be avoided, although some form of identification might be advisable, for example coloured badges;
- every member of the party should be informed in advance what to do if children become lost or separated from the party;
- a cheque-book with banker's card or emergency cash is useful;

- a small travelling first-aid kit must be carried (see Appendix 11 for recommended contents);
- drinking water should be taken;
- a bucket is useful for long journeys;
- regular toilet stops are best planned in advance;
- a mobile telephone is useful.

For camping/residential stays the following additional ideas are useful:

Food

Ensure sufficient is available and appropriate to the dietary need of children.

- Good practice in food handling should be maintained (see Chapter 8).

Means of cooking

Ensure equipment is safe and suitable, camping gas containers must be stored away from tents used in a ventilated space and turned off when not in use.

Drinking/washing

Ensure that water is from a safe source; if in doubt, do not use.

Fires

If fires are permitted there must be a properly established, recognized area for fire, with a hard, positively non-combustible base, which can be dampened down in periods of dry weather. Fires must be built at ground level. Fires dug into the ground are unacceptable. Fires may be built in metal containers, provided they are effectively stabilized. They must be sited away from areas of high activity and from tents and vegetation. A safety zone of 2 m around the fire must be out-of-bounds to children.

A supply of water, sand or loose soil adequate to extinguish the fire, and a fire blanket, must be kept near by, especially if it is a cooking fire. If such means are not available, then no fire may be built. Clean, cold water is also essential for immediate first-aid in case of accidents. Any fire area must be completely extinguished when no longer needed and not left to burn themselves out. A fire must not be left unattended until it the fire is completely out. The fire must be repeatedly doused with water and raked until it has been completely extinguished. Special care must be taken in the case of container fires, since a metal container will retain heat for a considerable time.

Children's personal welfare

A nominated key worker must be appointed for each child, who must pay particular attention to the welfare of that child.

General safety

Ensure that there are suitable toilet and washing arrangements, determine the location of all danger points near the location and warn children; contact a local doctor and enquire in advance as to whether it is possible to register all party members as temporary patients, determine the location of the nearest telephone, find out the times that the nearest hospital accident and emergency departments are open for all times of the day and all days of the week.

Sleeping arrangements

Sleeping bags and bedding must be dry and well aired and tents sufficiently water – and weather-proofed, sleeping areas must be clean and tidy, naked lights must not be used in tents. Adults must not sleep in the same tents as children.

10.9 VEHICLES

Many adventure playgrounds use minibuses to transport children on trips and outings. Although these vehicles have a good safety record, it is important to ensure that they are operated legally and as safely as possible.

Minibuses are vehicles which carry more than 8, but not more than 16, seated passengers in addition to the driver.

Full information on the operation of minibuses can be obtained from the Community Transport Association. The following summarizes the information provided by that organization.

10.9.1 Section 19 permits

Minibus drivers and operators have responsibility for the safety of passengers both under road traffic legislation and as part of their general duty of care.

Minibuses which are operated commercially for use by the general public are subject to stringent regulation concerning the qualifications of drivers and operators. In contrast there are no regulations at all in respect of minibuses where no charge whatsoever is made to any passenger.

However, many organizations operate minibuses for the benefit of their members and charge passengers in order to contribute to the running costs of the vehicles.

To ensure that these non-profit-making bodies operate their vehicles safely they are required to operate under a Section 19 permit.

These permits are issued by designated bodies such as local authorities. The following requirements must be met before a Section 19 permit can be issued:

1. The vehicle licensed must always be used by the organization to whom the permit has been granted, or affiliated/associate members of that organization.
2. The vehicle must not be used by members of the general public.
3. The vehicle must not be operated for profit, or for an activity which, in itself, is carried out for profit.
4. The vehicle must meet the requirements for initial fitness as defined by the Department of Transport regulations. These are detailed and complex. They include the provision of first-aid equipment and fire extinguishers. It is good practice when hiring or purchasing a minibus to ask the vendor to confirm in writing that the vehicle complies with the minibus construction requirements as set out in the Road Vehicles (Construction and Use) Regulations 1986. (Although not required by these Regulations, the Community Transport Association recommend that vehicles must be fitted with automatic fuel and automatic electric cut-off devices, and opt for an engine run on diesel, which is less combustible than petrol.)

Every vehicle operating under a Section 19 Permit must display a permit disc.

10.9.2 Insurance

Insurance is compulsory. Minibus operators must check insurance policies for the following:

- whether it covers all the uses to which the vehicle is put;
- whether the contents are covered;
- the total number of passengers allowed to be carried, and the total weight;
- whether there are any restrictions on drivers.

The insurance is unlikely to cover personal use, and drivers must be warned of this.

10.9.3 Maintenance and MOT

It is essential that minibuses are maintained in a fit and serviceable condition. Operators are advised to keep a log and records of servicing and maintenance. Permit vehicles are subject to spot checks on their roadworthiness.

Every minibus requires an MOT Certificate from its first 'birthday' and annually thereafter. If the capacity is 13 passenger seats or more it requires a Class V test, which can only be carried out at HGV testing stations. Advanced booking is usually necessary.

10.9.4 Drivers

Every minibus driver must:

- be at least 21 years of age and have a valid full licence but the insurance company may insist on additional requirements such as medical conditions or older minimum age;
- have received some training in minibus driving under the conditions in which they are likely to be driving, this must include familiarization with the vehicle, emergency procedures, on-the-road assessment, training in the use of safety equipment, and where appropriate, wheelchair securement and use of passenger lifts, and must be ongoing;
- know what their responsibilities are, including to obey the Highway Code, to check the vehicle's roadworthiness before setting off (see Appendix 15);
- drive for limited hours to avoid fatigue.

Drivers of minibuses operated under a Section 19 permit are not subject to drivers' hours regulations, unless they drive abroad (contact the Community Transport Association for Details). However they must not drive:

- when they feel tired or unwell, or when taking medication where driving is contra-indicated;
- when under the influence of drugs or alcohol;

or be expected to drive for a significant length of time after a day at work (including returning from a residential activity or trip away), for example take a break of at least 10–15 minutes after every two hours.

They must always carry a suitable form of identification e.g. driving licence.

PLAYLINK strongly advises that a police check be properly carried out on any persons working, or volunteering to work, as a minibus driver (or escort, see below) who will have substantial access to children. Where playgrounds do not have access to police vetting schemes, a rigorous recruitment, management awareness and supervision procedure can add safeguards. See section in Chapter 3.

10.9.5 Children as passengers

Another adult as supervisor/escort is essential when children are carried as passengers. That person must be inducted into what is expected of them and familiarised with the vehicle, including the emergency exits fire extinguisher and first-aid kit. At least one of either the driver or the escort must hold a first-aid certificate. Supervisors must be vetted in the same way as drivers (see above).

A useful checklist for drivers and escorts is provided at Appendix 15. The following list highlights some other issues concerned with children as passengers.

- Never use the 'three-for-two' rule which allows three children to count as two passengers, or seat three children in two seats.

- Never allow children to stand up whilst the vehicle is travelling.
- Children must avoid sitting at the very rear or very front sets if possible, as these are danger zones for front and rear collisions.
- Do not use side facing seats unless there is no alternative.
- For information on use of seat belts see below.

The Community Transport Association has devised a useful framework of rules for children using minibuses which is reproduced at Appendix 16.

10.9.6. Seat belts

Seat belts are required to be fitted in front passenger and driver seats of all minibuses, and these, and any other seat belts fitted, must be worn. There are special rules for children under three years and under four years of age.

In addition, from 10 February 1997, a new regulation of the Road Traffic Act 1988, (The Road Vehicles (Construction and Use) (Amendment) (No. 2) Regulations 1996) requires all minibuses (and coaches), where three or more children are carried as passengers, on an organized trip, or on a journey made for the purposes of a trip, to be fitted with seat belts, in sufficient number for each child in the party.

It is in addition good practice to:

- never allow children to share a seat belt;
- use booster and child seats where appropriate;
- never exceed the seating capacity of the vehicle – i.e. one passenger per seat.

More information on seating is obtainable from the Child Accident Prevention Trust, manufacturers, or the road safety officer of the local authority.

A Community Transport Association guide to good practice and legal standards concerning the fitting of seat belts was due for completion in 1996.

10.9.7 Luggage

Luggage must be secured and must not block gangways or exits. It must not overload the vehicle. It must be distributed correctly.

Roof racks change the handling characteristics of a vehicle and the driver must be prepared for this. The need for people to climb on top of vehicles or their loads must be avoided as far as possible. Where it is unavoidable, effective measures must be taken to prevent falls.

Trailers restrict the legal top speed (to 50 mph on single carriage ways and to 60 mph on dual carriageways and motorways) and access to the third lane on a three lane motorway. They also restrict the rear door access in an emergency and must not be used unless there is an additional emergency side door easily accessible to passengers in the rear.

10.9.8 Passenger lifts and ramps

Specific regulations are in force for minibuses which have passenger lifts and ramps, and for passengers in wheelchairs. These are available from the Community Transport Association. (See Appendix 17 for details.)

10.9.9 Emergencies

Every driver and escort/helper must know:

- what to do in the event of a breakdown or accident – including familiarity with procedures set out in the Highway Code, and the emergency contact procedure (see above);
- what to do in other emergencies;
- how to use the fire extinguisher(s);
- at least one of either the driver or the escort must know how to use the first-aid kit on the vehicle (this will require first-aid training).

Other emergency equipment must include:

- pen, paper and organizational details;
- insurance details;
- phonecard/change or mobile phone;
- webbing cutter;
- AA/RAC or other breakdown policy details and contact telephone number;
- red warning triangle.

10.10 LEGISLATIVE BACKGROUND

The following are the main pieces of legislation which relate to the provision of off-site activities organized from adventure playgrounds:

- The Health and Safety at Work etc. Act 1974, and its Regulations:
  - The Management of Health and Safety at Work Regulations 1992;
- The Transport Act 1985, and its Regulations:
  - The Road Vehicles (Construction and Use) (Amendment) (No. 2) Regulations 1996;
- The Children Act 1989;
- The Activity Centre (Young Persons' Safety) Act 1995 as set out in the Adventure Activities Licensing Regulations 1996.

10.11 FURTHER READING

The following are the main source documents for this chapter:
Buckinghamshire County Council (1985) *School Visit to Cornwall by Stoke*

Poges Middle School in May 1985, Report of the Chief Education Officer.

Community Transport Association (1994) T*he Operation of Minibuses in the Voluntary Sector.*

Community Transport Association (1990) *Your Minibus – is it Legal?*

Outdoor Adventure *Activity Providers Code of Practice*, 1994.

Guidance to the *Licensing Authority on the Adventure Activities Licensing Regulations 1996*, HSE.

PART TWO

SUMMARIES OF LEGISLATION AND
REGULATION

Chapter 11

Health and Safety at Work etc. Act 1974, the Regulations

11.1 THE REPORTING OF INJURIES, DISEASES AND DANGEROUS OCCURRENCES REGULATIONS 1985

11.1.1 Introduction

The Reporting of Injuries, Diseases and Dangerous Occurrences Regulations (RIDDOR) 1985 places three principle duties on employers:

1. the keeping of records of accidents at work – an accident book;
2. the reporting of accidents at work, including to those who are not employed at the place of work;
3. the reporting to the Health and Safety Executive of dangerous occurrences (near accidents).

Reporting is made to the appropriate enforcing authority for the premises. For adventure playgrounds operated by local authorities this is to the Health and Safety Executive, for others it is the Environmental Health Officer of the local authority.

The reporting of accidents is related to the extent and nature of the injury suffered. Reportable accidents relevant to adventure playgrounds include:

- fractures of the skull, spine or pelvis;
- fractures of the arm, wrist, leg or ankle;
- amputation of the hand, foot, finger, thumb, toe or any part thereof;
- loss of sight of eye, penetration injury to the eye, chemical or hot-metal burn to eye;
- injury (including burns) requiring immediate medical treatment, or loss of

consciousness because of electrical shock;
- loss of consciousness due to lack of oxygen;
- acute illness/loss of consciousness resulting from absorbtion of any substance by inhalation, ingestion or through the skin;
- any other injury which results in admission to hospital for more than 24 hours;
- for employees only, absences due to injury at work, where these absences are for more than three days;
- fatalities;
- the reporting the Health and Safety Executive of deaths as a result of a reportable injury within one year.

All fatal accidents, major injuries and dangerous occurrences have to be reported immediately (e.g. by telephone).
Reportable occurrences could include:

- failure or overturning of lifting equipment or excavators;
- collapse of building structures of more than 5 tonnes in weight;
- collapse of floor or wall of any building used as a place of work.

A list of Health and Safety Executive Offices is provided at Appendix 2.

11.2 THE CONTROL OF SUBSTANCES HAZARDOUS TO HEALTH REGULATIONS 1994 (COSHH)

11.2.1 Introduction

The COSHH Regulations impose a duty on employers in respect of employees and others affected by their work. To a lesser extent it also imposes duties on employees. The foreseeably principal duty is to prevent exposure of employees and others to substances hazardous to health or, if prevention is not attainable, to control exposure adequately.

They provide a basic system for managing risks. This system is based on carrying out a risk assessment.

11.2.2 Assessment of risk

If the risk assessment (see above) carried out in compliance with the Management of Health and Safety at Work Regulations has identified hazardous substances in the workplace and potential risks associated with them, then a more detailed risk assessment will be required. For the purposes of COSHH, hazardous substances include:

- those listed as toxic, very toxic, harmful, irritant or corrosive;
- substances with occupational exposure limits;
- harmful micro-organisms;

- dusts of any kind in substantial concentrations;
- any other substances creating comparable hazards to health, such as pesticides.

These can be recognized by experience, from information provided by suppliers, including product data sheets, and by asking advice. Some Health and Safety Executive Guidance Notes provide specific information on dangerous substances.

At an adventure playground these could include domestic cleaning materials, paints, solvents, adhesives and pesticides.

Once identified it is necessary to undertake a detailed risk assessment in order to establish how the substances can become a hazard. This will require looking at how and where they are stored and used, who may be affected, how they would be affected (e.g. swallowing, breathing in etc.), and what measures are currently being taken to prevent or control exposure.

If as a result of this assessment it has been concluded that there is a risk to health then, in order to comply with the regulations, it will be necessary to:

- prevent exposure by changing the process, replacing it with a safer alternative or using it in a safer form;
- or, if prevention is not reasonably practicable, adequate control of exposure by a combination of the following: total enclosure of the process, restricted exposure, general ventilation and using systems of work which minimise chances of spills.

Only if the exposure cannot be adequately controlled must personal protective equipment be provided to control the hazard.

Once an assessment has been made it must be recorded, and reviewed regularly.

A risk assessment for the purposes of COSHH at an adventure playground can be broken down into five stages.

Stage 1   Identification of the substances used, what they are used for and any hazardous raw materials they contain or hazardous materials they produce.

Stage 2   Collection of knowledge about the substances, including any hazards and risks.

Stage 3   Assessment of risks by relating current uses and the controls in place, to the likely causes of exposure and who may be exposed.

Stage 4   Identification of ways of controlling the hazards and risks identified in Stage 3, including safe operating procedures, maintenance systems, and personal protective equipment required.

Stage 5   Monitoring the effectiveness of the ways used to control the hazards and risks.

11.3 WORKPLACE (HEALTH, SAFETY AND WELFARE) REGULATIONS 1992

11.3.1 Introduction

The Health and Safety at Work etc. Act requires that any person who has, to any extent, control of premises or of access to and from them, or of plant or substances within them, shall so far as is reasonably practicable ensure that they are safe and without risk.

The Act places a duty upon employers only in relation to workplaces where any of their employees work. They do not expand the general duties towards persons who are not employees which are contained within the Health and Safety at Work etc. Act. They require employers to ensure that adequate welfare facilities are provided for people at work.

The Regulations affect all workplaces which are used for the first time after 31 December 1992, and modifications, extensions and conversions started after that date, and existing workplaces on 1 January 1996.

Employers have a duty to ensure that the workplaces under their control comply with these Regulations, and that any facilities required by the Regulations are properly provided.

Where, to some extent, another person controls part of a workplace (e.g. shared facilities, means of access etc.) then they will also have responsibilities under these Regulations.

Guidance to these Regulations recommends that the views of employees be sought.

The following summarizes the main components of these Regulations which are relevant to adventure playgrounds. For full details, reference to the appropriate Health and Safety publication will be required from HSE Publications.

11.3.2 Equipment, devices and systems

These Regulations apply to equipment, devices and systems, a fault in which is liable to result in a failure to comply with any of the Regulations.

Such items must be maintained in efficient working order and in good repair. 'Efficient' in this context means from the point of view of health and safety. Any defects must be put right and steps taken to protect anyone who may be at risk as a consequence of the defect. Steps must be taken to ensure that the repair and maintenance are carried out properly.

A system of maintenance is required for certain equipment, for example, emergency lighting and fencing. This system must specify what action is required and how often.

11.3.3 Temperature in indoor workplaces

The Regulations set out requirements concerning the temperature of indoor workplaces, and methods of heating and cooling. They also require the provision of sufficient thermometers to assess temperature levels.

Temperature levels should provide reasonable comfort without the need for special clothing. Where severe physical effort is not required, this will normally be at least 16 °C. Where the temperature is consistently high reasonable steps should be taken to moderate it. Where a reasonably comfortable temperature cannot be achieved throughout a workroom, local heating/cooling should be provided. Where this is not achievable, suitable protective clothing and rest facilities must be provided.

Heating systems should not discharge the products of combustion into the workplace.

11.3.4 Lighting

The Regulations require that every workplace should have suitable and sufficient lighting. Windows and skylights should be clean and free from obstructions. Fire precautions may require the lighting of escape routes.

11.3.5 Cleanliness and waste materials

This section is concerned with the furniture, furnishings, floor surfaces, walls and ceilings and the disposal of waste materials.

In all cases these items are required to be kept sufficiently clean. The level of cleanliness will depend upon the use to which the workplace is put. Cleaning should be carried out as necessary, and a suitable method employed which does not endanger health and safety.

High standards of cleanliness will be required for the purpose of infection control in areas where food is handled.

Waste materials should not be allowed to accumulate in the workplace, except in appropriate receptacles.

11.3.6 Condition of floors and traffic routes

This section is concerned with the condition of floors and traffic routes. It requires that these should be of sound construction, free from obstructions, holes, slopes, uneven and/or slippery surfaces which are likely to cause a slip, trip or fall, drop or loss of control of a carried item, or cause instability or loss of control of vehicles and/or their loads.

Staircases should have a secure fence on every open side and a substantial handrail on at least one side.

11.3.7 Falls or falling objects

All reasonable steps must be taken to prevent falls by a person, or a person being struck by falling items.

Guidance on this includes the provision of secure fencing where a fall of over 2 m is possible, or where a fall of less than 2 m is likely to result in serious injury. Fencing should be sufficiently high and infilled, and of appropriate strength.

Fixed ladders should not be provided in circumstances where it would be practical to install a staircase. Where they are unavoidable they should be of sound construction, properly maintained and securely fixed.

Materials and objects should be stored and stacked in such a way that they are not likely to fall and cause injury. Racking should be of adequate strength and stability.

The need for people to climb on top of vehicles or their loads should be avoided as far as possible. Where it is unavoidable, effective measures should be taken to prevent falls.

11.3.8 Windows, and transparent or translucent doors, gates and walls

There are Regulations concerning transparent or translucent materials. Where these are found in doors and gates below shoulder height, and in windows, walls and partitions at waist level or below, they should be of safety material, or be adequately protected against breakage.

These materials should be marked where necessary to make them apparent.

Windows, skylights etc should be capable of being opened and of being cleaned without exposing the operator to risk. When open they should not project into an area where persons are likely to collide with them.

11.3.9 Sanitary conveniences, washing facilities and drinking water

Suitable and sufficient sanitary conveniences must be provided in the workplace.

Washing facilities should be suitable and sufficient, including showers if required for health reasons, or if work is particularly strenuous or dirty.

An adequate supply of wholesome drinking water should be provided for all persons, with a sufficient number of suitable cups or drinking vessels, unless it is a jet supply system (drinking fountain).

11.3.10 Lockers, changing and rest/meals facilities

These Regulations require employers to provide accommodation for storing of workplace clothing and changes of clothing, and facilities for changing.

They also require employers to provide suitable rest and meal facilities, including seating and facilities for preparing food or heating their own food.

These facilities must be kept to a suitable hygiene standard.

Rest areas should be so arranged as to enable employees to use them without experiencing discomfort from tobacco smoke.

11.4 PERSONAL PROTECTIVE EQUIPMENT AT WORK REGULATIONS 1992

11.4.1 Introduction

These Regulations concern the provision of suitable Personal Protective Equipment (PPE). They only apply to people who are employees, and not to others such as voluntary workers. However the provision of PPE to volunteers may be required in order to comply with the general duty under the Health and Safety at Work etc. Act, Section 3.

PPE is to be considered as the last resort in the prevention of risks to employees. It should therefore only be provided where a risk to health or safety cannot otherwise be adequately controlled.

PPE must be readily available, and free of charge to the employee.

Selection of PPE requires an assessment of the risk and the capabilities of the equipment. It should take account of ergonomic and other factors, which are best decided in consultation with the employees.

Employers must ensure that any PPE they purchase complies with the UK legislation. All PPE which has been approved by the HSE bears the European CE mark.

Where a mixture of elements of PPE are worn simultaneously, items should be compatible with each other.

11.4.2 Assessment, maintenance and storage of PPE

The selection of PPE requires an assessment process to be undertaken. This involves:

• identifying the potential hazards and the parts of the body endangered;
• collection of information from manufacturers concerning applications of PPE;
• selection of PPE suitable for the application, taking into account the nature of the job and the demands it places on the employee.

Selection is part of a continuing process concerned with the proper use and maintenance of the equipment and the training and supervision of users.

An appropriate system for maintenance of PPE is required. This will include levels of responsibilities, stocking of spare parts, and the following of manufacturers' maintenance schedules and instructions.

PPE should be appropriately stored.

11.4.3 Information, instruction, training and use

Employers are required to provide such information, instruction and training as is adequate and appropriate in order for the user to know:

• what risk the PPE will avoid or limit;
• the purpose for which the PPE is being used;
• any action required of the user to ensure that the PPE continues to function properly.

This will include training in the proper use of PPE, its correct fitting and wearing, and include elements of theory as well as practice.

The extent and frequency of training required will depend upon the type of equipment, how frequently it is used and the needs of the people being trained. Refresher training may be required.

Employers also have a duty to ensure that PPE is properly used. As well as the training of users, this also requires the issuing of instruction covering use and storage of the equipment.

Arrangements should be put in hand to ensure that employees can report the loss of, or defects in, PPE.

11.5 PROVISION AND USE OF WORK EQUIPMENT REGULATIONS 1992

11.5.1 Introduction

The primary objective of the Provision and Use of Work Equipment Regulations (PUWER), is to ensure the provision of safe work equipment and its safe use. PUWER amplifies and makes more explicit the general duties on employers to provide safe plant and equipment.

It overlaps with many other requirements and should not be considered in isolation. Many of the protective and preventative measures contained within these Regulations will have been identified as part of the risk assessment required by the Management of Health and Safety at Work Regulations 1992.

Work equipment is defined very widely. It is easier to list those items which are not work equipment, which are livestock, substances (e.g. water), structural items (e.g. walls) and private cars.

The prime duty for ensuring health and safety rests with employers. However employees have duties too. Where employees are using work equipment which has been properly provided and who have received the necessary training and appropriate instructions, they are then required to use that equipment correctly.

The Regulations in general apply to all work equipment provided for use from 1 January 1993. This includes second hand-equipment and hired and/or leased equipment brought into use after this date.

11.5.2 Suitability of equipment

Every employer is required to ensure that work equipment is so constructed or adapted as to be suitable for the purpose for which it is provided. This requires an assessment of the working conditions and the risks to health and safety to the person carrying out the work. In addition, employers are required to ensure that work equipment is only used for the operations for which it is suitable.

This selection will take place within the risk assessment carried out under the Management of Health and Safety at Work Regulations. The selection involves an assessment of:

- initial integrity of the equipment;
- the place where it is to be used;
- the purpose for which it will be used.

11.5.3 Maintenance of equipment

It is the responsibility of employers to ensure that work equipment is maintained in efficient working order and in good repair. 'Efficient' in this context relates to how the condition of the equipment might affect health and safety.

Routine maintenance includes periodic lubrication, inspection and testing, based on the recommendations of the manufacturer. Other legislation may require specific maintenance.

Planned preventative maintenance may be necessary when inadequate maintenance could cause the equipment to fail in a dangerous way. This can be achieved through a system of written instructions and procedures.

Although there is no requirement for maintenance logs to be kept, it is recommended that a record of maintenance be kept. Any log should be kept up to date.

11.5.4 Information, instructions and training

Employees should be provided with adequate information and, where appropriate, written instructions on the use of work equipment. This information should include:

- the conditions in which and the methods by which the work equipment may be used;
- foreseeable abnormal situations and the action to be taken if such conditions were to occur;
- any conclusions to be drawn from experience in using the work equipment.

Information should be readily comprehensible to those concerned. The information conferred is primarily that provided by the manufacturers or suppliers of the equipment, who have duties to provide relevant information. It should cover all the health and safety aspects of use that will arise and any limitations/special circumstances.

In addition to providing information, employers are required to provide adequate training to those using work equipment, or supervising those who do so.

11.5.5 Conformity with European Union requirements

This Regulation also aims to ensure that when work equipment is first provided for use that it meets certain health and safety requirements. It places this duty on employers.

Where the item of equipment to be provided is covered by an EU Product Directive, employers must obtain a copy of the EU Declaration of Conformity from the manufacturer or supplier.

Equipment which is not covered by a Product Directive, including all secondhand equipment, must comply with the other requirements of these Regulations (see below) when first taken into use. In case of doubt, advice can be obtained from the local Health and Safety Executive Office. (See Appendix 2.)

11.5.6 Other requirements

There are a number of other detailed requirements contained within these Regulations. These are summarized below:

* taking account of dangerous parts of machinery (Reg. 11);
* protection against specific hazards – deals with the prevention and/or control of hazards as a result of ejections from or disintegration/explosion of work equipment, abrasive wheels are specifically referred to (Reg. 12);
* protection from burns or scalds as a result of high or low temperatures related to work equipment (Reg. 13);
* the provision of control systems, including stop controls (Regs. 14–18);
* isolation from energy sources to ensure that inadvertent reconnection is not possible (Reg. 19);
* ensuring that the many types of work equipment which may fall over, collapse or overturn are appropriately stabilised (Reg. 20);
* the provision of suitable and sufficient lighting for the use of work equipment (Reg. 21);
* the construction of work equipment in such a way that it can be maintained without risk to health and safety (Reg. 22).

Regulations on markings and warnings are particularly relevant to adventure playgrounds.

These are concerned with the provision of clearly visible, relevant markings for work equipment giving instructions and/or warnings. They should be easily perceived, understood and unambiguous.

11.6 MANUAL HANDLING OPERATIONS REGULATIONS 1992

11.6.1 Introduction

These Regulations seek to prevent injury which may occur in the process of manual handling. Where a risk assessment indicates the possibility of risk to employees from the manual handling of loads the requirements of these Regulations should be observed.

The Regulations set out a hierarchy of measures:

- avoid manual handling as far as is reasonably practicable;
- assess any hazardous manual handling operations that cannot be avoided;
- reduce the risk of injury so far as is reasonably practicable.

Once steps have been taken to reduce the risk of injury through manual handling they should be monitored to check that they are effective and continue to be so.

11.6.2 Assessment of risk

Where risk assessment has indicated that the avoidance of manual handling is not reasonably practicable, then it is recommended that a more specific assessment should be carried out.

In most cases this assessment can be carried out by the employers themselves. Areas of knowledge likely to be relevant to the successful assessment of risks are:

- the requirements of these Regulations;
- the nature of the handling operations;
- a basic understanding of human capabilities;
- identification of high-risk activities;
- practical steps to reduce risk.

The views of staff can be of particular importance in this assessment.

An assessment checklist is set out as an appendix to the Regulations, which also contain useful information and guidance on manual handling. That checklist is provided in this document at Appendix 18.

Once an assessment has been carried out and appropriate systems of work are laid down, it is the responsibility of employees to follow these systems.

11.7 ELECTRICITY AT WORK REGULATIONS 1989

11.7.1 Introduction

The Electricity at Work Regulations 1989 aim to ensure that electrical equipment is constructed or protected and used as far as is reasonably practicable to prevent risks to health and safety.

In February 1990, Guidance Note GS23 (Electrical Safety in Schools) was issued by the Health and Safety Executive, taking into account the requirements of these regulations with respect to schools. This document has been used as a source document for the following section because of similarities in the environments of schools and play facilities.

11.7.2 Competent persons

The regulations require that only qualified, competent personnel shall specify equipment, design systems, and install or carry out work on electrical equipment. As carrying out work on electrical equipment can range from changing a plug to installing high-voltage systems, the degree of competence required will vary. The following guide is provided to assist in decisions on appropriate levels of competence:

Level 1

Competent to – judge the competence of others, advise on policy matters, oversee and design selection, installation and maintenance of electrical system.
Qualification – a Chartered Incorporated Electrical Building Services Engineer.

Level 2

Competent to – design and specify electrical installations and systems under supervision, oversee, test and accept electric works, supervise electrical operations.
Qualification – an Electrical Technician or Technical Engineer.

Level 3

Competent to – interpret drawings and carry out installations of electrical equipment, test electrical installations and equipment, maintain and diagnose faults on installations and equipment.
Qualification – JIB registered Approved Electrician or Electrician, or a trainee Electrician working directly under the supervision of an Electrician.

Level 4

Competent to – fit a 13 amp plug to a fixed or portable appliance, replace lighting lamps/bulbs and ensure safe disposal, isolate, disconnect and reconnect single-phase electrical equipment and appliances.
Qualification – a person who has received training and is competent to carry out minor specified tasks in a safe manner.

11.7.3 General approach

Fixed electrical installations (ring mains etc.), provided they have been installed in accordance with the Institution of Electrical Engineers' Wiring Regulations are generally held to be safe providing that they are maintained and inspected every five years. This may need to be more frequent if there is a suspicion of damage or abuse. An inspection/test certificate should be obtained.

Where fixed electrical installations are likely to be modified/extended by persons with limited electrical knowledge, then such modifications should only be undertaken with the consent of a competent person.

All electrical equipment, including socket outlets and fittings such as lights, radiant heaters etc. should be chosen according the use they will be subjected to.

11.7.4 Apparatus

Most electrical apparatus used in adventure playgrounds will be of domestic or commercial pattern. It is recommended that an inventory be compiled of all equipment, and that a competent person assesses all items.

All portable equipment should be routinely inspected and tested. A register or log book should be kept for each item.

All Class 1 (which are earthed through the lead to the ringmain) hand held portable equipment such as drills should be subject to a detailed inspection and test by a competent person at least every 12 months, including an earth continuity test. An appendix to Guidance Note GS23 provides a useful checklist for these inspections.

All Class II (double insulated, which do not have an earth lead to the ringmain) should also be subject to a detailed inspection and test as above. In addition they should be visually tested before each use.

Audio-visual and other equipment with exposed metalwork should be treated as a Class I item, unless the manufacturer specifically claims that it is double insulated.

Flexible cables should be selected, maintained and used so that there is adequate protection against foreseeable mechanical damage.

11.7.5 The use of residual current devices

A higher standard of electrical protection can be achieved through the use of Residual Current Devices (RCDs). Where electrically operated hand-held portable equipment is used outdoors the source of supply should be controlled by an RCD. If used they should be tested frequently by means of the test button on the unit and, when the installation is routinely tested the tripping current and timing of RCDs should be checked.

The use of RCDs is required as back up protection against electrical shocks in all cases where the electrical equipment is used outdoors. This equipment should be of low-voltage type (see below). There is also a requirement that plugs and sockets used outdoors comply with BS 4343 as standard 13–A plugs and sockets are not proof against moisture, water or dirt.

Health and Safety Executive Guidance Note PM 32 gives full details of the use of RCDs with portable electrical equipment.

11.7.6 Lighting and sound supplies

In Guidance Note GS 50 (Electrical Safety in Places of Entertainment) published by the Health and Safety Executive in May 1991, there are some useful guidelines on the electrical safety of lighting and sound supplies.

These have been drawn up in the context of a number of accidents in which entertainers have received electric shock from their equipment, some of which have been fatal.

They are of relevance to adventure playgrounds which provide for entertainments using these media, and it is recommended that GS 50 is obtained as a reference document.

11.7.7 Low voltage equipment

Where possible, electrical equipment used at an adventure playground should operate at 110 volts and be used with a 110 volt transformer. A residual current device installed close to the point of supply is recommended for all uses and is required for work outdoors. Plugs and sockets used outdoors should comply with BS 4343 (male/female fittings). Electrical tools must not be used in damp or wet conditions or in the presence of flammable vapours or gases.

11.8 GAS SAFETY (INSTALLATION AND USE) REGULATIONS 1994

These Regulations are concerned with the safety of gas fittings, meter installations, pipework and gas appliances. They cover liquid petroleum gas (LPG) as well as natural gas.

There is a general requirement that all persons who carry out work on gas systems should be approved by the Health and Safety Executive (CORGI registered). Employers are required to take reasonable steps to ensure that any work carried out on gas appliances at a place of work is undertaken by a CORGI registered person.

There are requirements concerning meters and regulators, installation pipework and gas appliances.

One important aspect of these is the requirement for maintenance of gas appliances. Employers are required to ensure that any gas appliance or installation pipework at a place of work under the employer's control is maintained in a safe condition to prevent risk of injury.

11.9 CONSTRUCTION (DESIGN AND MANAGEMENT) REGULATIONS 1994

These Regulations are relevant to those adventure playground providers who are involved in large construction projects. For example the design and construction of a new play building and site.

The Regulations place duties on all those involved in a construction project (clients, planning supervisors, designers and contractors) to ensure that all the health and safety responsibilities in respect of the project are properly discharged.

Those persons for whom a project is carried out (clients) must be reasonably satisfied that they only use competent people as planning supervisors, designers and principal contractor and be satisfied that sufficient resources, including time, have been or will be allocated to enable the project to be carried out in compliance with health and safety law.

Small construction projects are exempt from the bulk of these Regulations. These are those which will take no longer than 30 days or 500 person hours (which are not notifiable under the Regulations), and those on which the largest number of persons at work at any one time carrying out the construction work is less than five.

For all other projects a range of duties and responsibilities are required of the parties to the construction, and these are fully described in the Regulations.

Chapter 12

Other significant legislation

12.1 THE OCCUPIERS' LIABILITY ACTS OF 1957 AND 1984

12.1 Introduction

The Occupiers' Liability Act of 1957 codified the common law duty of care owed by occupiers of premises to all lawful visitors.

In 1984 a second Occupiers' Liability Act extended this duty of care to include trespassers who may be at risk of injury on the premises. This duty can be discharged by issuing of explicit warnings of the hazards concerned, for example by posting notices warning of the hazards. As children are considered to be less careful than adults there is a greater duty of care owed to them, and warnings which may be adequate for adults may not be so for a child.

'Premises' can include both the indoor and outdoor areas of an adventure playground.

12.2 THE OFFICES, SHOPS AND RAILWAY PREMISES ACT 1963

This legislation requires employers of staff working in offices, shops and railway premises to provide suitably ventilated, heated and clean working conditions.

It relates to adventure playgrounds only in respect of the parts of buildings where administration takes place.

12.3 UNFAIR CONTRACT TERMS ACT 1971

12.3.1 Introduction

The Unfair Contract Terms Act is concerned with ensuring that no liability to pay compensation for negligence can be avoided through contractual terms or otherwise.

Under the Act no contractual term or notice can exclude, or restrict, liability for death or personal injury resulting from negligence.

However a contractual term or notice can exclude or restrict liability for loss or damage due to negligence, if the term or notice is reasonable. In this context, reasonable means; fair and reasonable to allow reliance on it, having regard to all the circumstances obtaining when the liability arose or (but for the notice) would have arisen.

This is most relevant to adventure playgrounds in respect of any notices or disclaimers that aim to restrict or exclude the liability of the operators. For example on consent forms, notices and signs which seek to impose limits on the responsibility (e.g. 'we will not accept responsibility for your child if she wanders off while she is with us on the outing'). The Act limits the extent to which operators can avoid liability by the use of such notices. However their use as a source of information on health and safety matters is still valid.

12.4 FOOD AND ENVIRONMENTAL PROTECTION ACT 1989 PART III – THE CONTROL OF PESTICIDES REGULATIONS

12.4.1 Introduction

These Regulations are concerned with the storage and use of pesticides. Pesticides are substances used for protecting plants and killing bugs etc.

As pesticides are also likely to be identified as hazardous substances in a general risk assessment they would also be subject to assessment, under the COSHH Regulations.

These Regulations are concerned primarily with steps to be taken to ensure that humans, creatures, plants, and the environment, are not adversely affected as a result of storage and use of pesticides.

12.4.2 Storage and use

Any person who stores and/or uses pesticides must take all reasonable precautions to protect the health of human beings, creatures and plants and to safeguard the environment and, in particular, to avoid the pollution of water, and users should be competent in the duties which they are called upon to perform.

No person shall use pesticides unless that person has received adequate instructions and guidance in this work, and is competent for the duties which the person is called upon to perform.

Pesticides may be present on adventure playgrounds as timber preservatives. Weed killers may also be present (these are herbicides, but similar care in use is required) and specific advice and guidance should be sought on their storage and use. Suppliers are required to provide this information. The use of timber preservatives is discussed in Chapter 6.

12.5 THE CONSUMER PROTECTION ACT 1987

12.5.1 Introduction

The Consumer Protection Act 1987 is concerned with providing redress in respect of damage which arises as a consequence of defects in products.

For the purposes of this Act, damage includes death or injury or loss of or damage to, any property; products include any goods; the producer is the person who manufacturers the goods; and a defect of safety arises where the product is not such as persons generally are entitled to expect.

Unlike the common law of negligence, where a case can only be made if negligence is shown, and the law of contract where a claim can only be made if a contractual relationship can be shown, this Act operates under the principle that legal proceedings can arise if damage occurs wholly, or partly by a defect.

In the context of adventure playgrounds the construction of any item for use on the site or in the premises is likely to be a product under the terms of this Act.

In 1994 a Statutory Instrument 'The General Product Safety Regulations' set out a new duty on producers; that they shall not place a product on the market unless the product is a safe product.

These Regulations apply to products supplied in the course of a commercial activity, whether for consideration or not. A safe product means any product which, under normal or reasonably foreseeable conditions of use, does not present any risk or only the minimum risk compatible with the products use.

12.6 FIRE PRECAUTIONS ACT 1971, AMENDED BY THE FIRE SAFETY AND SAFETY OF PLACES OF SPORT ACT 1987

12.6.1 Introduction

These Acts relate to premises which are licensed for music and dancing. They form the legislative background for considerations of fire safety, for example those made as part of registration of premises under the Children Act 1989.

12.6.2 Fire procedures

The requirements in respect of fire procedures are:

- the fire brigade should be called immediately, irrespective of the size of the fire;
- one person (but see above) to be given responsibility for ensuring that pre-planned action is carried out, and who should meet the fire brigade upon arrival;
- all procedures should take account of possible absence of personnel from premises, eg because of sickness, or leave etc.

12.6.3 Fire equipment

All persons should be trained to use fire extinguishers. The following types of fire extinguisher are available:

Type	Colour	Suitable for
Water	Red	Ordinary combustible fires for example wood and paper, but not suitable for flammable liquids
Foam	Cream	Small liquid spill fires
Dry powder	Blue	Flammable liquid fires
Carbon dioxide	Black	Fires involving electrical equipment

The green vaporizing Liquid (Bcf) extinguishers are no longer recommended as they are environmentally damaging and give off toxic fumes.

The type of extinguishers provided should be suitable for the risks involved. They should be clearly marked with an information sign provided alongside. They should be adequately maintained and a record should be kept of all maintenance, inspections and tests.

12.6.4 Fire alarm

A manually operated fire alarm is required in licensed premises.

12.6.5 Means of escape

In certain premises a 'means of escape' certificate/fire certificate should be available on demand. For all premises the following apply:

- all doors should work;
- all fire doors should be indicated;
- sliding doors should be marked with direction of movement;
- all doors should be marked and unlocked and gangways and escape routes should be clear.

12.6.6 Good housekeeping

The following good housekeeping is recommended:

- provide proper ashtrays in any areas where smoking is permitted
- keep combustible waste in closed bins
- combustible materials should be removed daily or locked away
- rubbish should not be left in building
- no-smoking areas should be enforced.

12.6.7 Fire drill

There is no statutory requirement that a fire drill be undertaken, but it is good practice to do so and may be a requirement of Children Act registration.

12.6.8 Emergency lighting

The provision of emergency lighting is a requirement of the licensing of buildings for music and dancing and may be a requirement of Children Act registration.

12.7 THE FURNITURE AND FURNISHINGS (FIRE) (SAFETY) REGULATIONS 1988 (AS AMENDED IN 1989)

These Regulations prescribe tests for materials used in the construction of furniture and to notify the consumer of the results of those tests by means of a visible label. These labels are of two types:

1. 'Resistant' – a square label coloured white with a green border and the word resistant in black. This indicates that furniture has passed tests for fire-resistant filling material, cigarette resistant upholstery and match resistant cover fabrics.
2. 'Warning' – a triangular label coloured white with a red border. This indicates that the furniture has passed the test for fire-resistant filling material and cigarette resistant upholstery only.

12.8 THE FOOD SAFETY ACT 1990 AND FOOD HYGIENE REGULATIONS 1991

The Food Safety Act 1990 and Food Hygiene Regulations 1991 cover a broad range of activities relating to food. Food in this context includes ingredients and drink.

The Act sets out food safety requirements. These are that food must not have been rendered injurious to health, be unfit or so contaminated that it would be unreasonable to expect it to be eaten. Considerations include fitness of premises and handling of food. It places responsibilities on anyone who handles or prepares food for public consumption. The Regulations set standards for safe handling and preparation of food and introduce training and registration requirements.

The Act is enforced by both local and central government. Adventure playground issues will be dealt with through the Environmental Health Department of the local authority and/or as part of the registration of day care facilities under the Children Act.

Those sites where prepared food is provided may be required to register with the local authority, and staff involved to have undertaken food handling training.

12.9 DATA PROTECTION ACT 1984

The Data Protection Act 1984 is concerned with the collection, holding and use of information on computers. Its principles are as follows:

- information should be obtained and processed lawfully;
- personal data must only be held for one or more specified lawful purpose;
- personal data can only be used for the purposes specified;
- personal data must be adequate, relevant and not excessive for the purposes specified;
- personal data must be accurate and up-to-date;
- personal data must not be kept longer than is necessary for the specified purposes;
- appropriate security measures must be in place to prevent unauthorised access to personal data.

Organizations holding personal data are required to register under the Data Protection Act.

Individuals who have data recorded in these ways are entitled, without undue delay or expense, to:

- be informed by the data user that their personal data is held;
- have access to the data held.

12.10 ENVIRONMENTAL PROTECTION ACT 1990, PART II

The Environmental Protection Act 1990 is principally concerned with the avoidance of air pollution. It is a complex and detailed piece of legislation, which impacts principally on adventure playgrounds in respect of fires on site, and the disposal of rubbish other than that collected as domestic rubbish.

Part II of the Act applies to the disposal of all controlled waste, which includes all household, industrial and commercial waste. It also covers special waste, i.e. waste which is difficult or dangerous to dispose of.

A legal duty of care is imposed on anyone who produces, carries or disposes of waste, to ensure that it is disposed of without causing pollution. One of the requirements included to ensure that this is carried out is that, when waste is transferred, it is accompanied by a full written description. This does not apply to domestic rubbish.

The Act also requires waste-collection authorities to draw up plans for recycling household and industrial waste.

12.11 SAFETY SIGN REGULATIONS 1980

These Regulations make it compulsory that any safety sign giving health and safety information or instructions to persons at work must comply with BS 5378 Part 1. This Standard relates to colour and style of signs.

12.12 THE ACTIVITY CENTRE (YOUNG PERSON'S SAFETY) ACT 1995, AS SET OUT IN THE ADVENTURE ACTIVITIES LICENSING REGULATIONS 1996

These Regulations are concerned with the licensing of outdoor activity centres to give assurance that good safety management practice is being followed.

The standards of adventure activity centres are to become regulated from October 1997. The Regulations only apply to activity centres which are run in return for payment. Registration is concerned with the number and qualification of instructors, the provision to the participants of information on safety, the provision and maintenance of proper equipment, and appropriate first-aid provision and medical back-up.

The Regulations are concerned with centres which provide any or all of the following activities:

- caving (underground exploration in natural caves and mines, including potholing, cave diving and mine exploration);
- climbing (climbing, traversing, abseiling and scrambling activities except on purpose-designed climbing walls or abseiling towers);
- trekking (walking, running, pony trekking, mountain biking, off-piste skiing and related activities when done in moor or mountain country which is remote, i.e. over 30 minutes travelling time from the nearest road or refuge);
- watersports (canoeing, rafting, sailing and related activities when done on the sea, tidal waters or larger non-placid inland waters).

Please note: the checklists and model forms in these appendices are designed so that they can be photocopied and used by play organizations. They may be copied free of restriction and without charge.

Appendix 1

List of contributors

PLAYLINK is grateful to the following individuals and organisations who have contributed to this publication:

Bob Hughes, Bridget Hanscomb, Chris Anderson, Danny Meeham, David Perkins, Donne Bucke, Dr Hugh Jackson, Gill Hinton, Gill Gibson, Jane Hibbard, Jess Milne, Joan Fisher, Ken Abbott, Mark Humphreys, Mary Januarius, Mick Conway, Mike Nussbaum, Paul Gallagher, Pete Mulvey, Peter Rawlinson, Peter Heseltine, Himadri Potter, Steve Derby, Sybil Lunn, Tony Chilton, Community Transport Association, Health and Safety Executive, Newham Council, Outdoor Adventure Activity Providers Association, PGL Adventure, Peterborough City Council, Portsmouth City Council, Southampton City Council, Hilary Koe.

The author gives special thanks to Donne Bucke for his patient and painstaking proofreading.

Appendix 2

List of Health and Safety Executive offices

Inner City House
Michell Lane
Victoria Street
Bristol
BS1 6AN

Tel: 0117 9290681

Covers: *Avon, Cornwall, Devon, Gloucestershire, Somerset, Isles of Scilly*

Priestley House
Priestley Road
Basingstoke
RG24 9NW

Tel: 01256 404000

Covers: *Berkshire, Dorset, Hampshire, Isle of Wight, Wiltshire*

3 East Grinstead House
London Road
East Grinstead
RH19 1RR

Tel: 01342 326922,

and

Maritime House
1 Linton Road
Barking
IG11 8HF

Tel: 0181 594 5522

Covers: *Barking and Dagenham, Barnet, Brent, Camden, Ealing, Enfield, Hackney, Haringey, Harrow, Havering, Islington, Newham, Redbridge, Tower Hamlets, Waltham Forest*

1 Long Lane
London
SE1 4PG

Tel: 0171 407 8911

Covers: *Bexley, Bromley, City of London, Croydon, Greenwich, Hammersmith & Fulham, Hillingdon, Hounslow, Kensington & Chelsea, Kingston, Lambeth, Lewisham, Merton, Richmond, Southwark, Sutton, Wandsworth, Westminster*

39 Baddow Road
Chelmsford
CM2 OHL

Tel: 01245 284661

Covers: *Essex (except the London Boroughs in Essex covered by the Barking Office), Norfolk, Suffolk*

14 Cardiff Road
Luton
LU1 1PP

Tel: 01582 34121

Covers: *Bedfordshire, Buckinghamshire, Cambridgeshire, Hertfordshire*

Belgrave House
1 Greyfriars
Northampton
NN1 2BS

Tel: 01604 21233

Covers: *Leicestershire, Northamptonshire, Oxfordshire, Warwickshire*

McLaren Building
2 Masshouse Circus
Queensway
Birmingham
B4 7NP

Tel: 0121 200 2299

Covers: *Birmingham, Coventry, Dudley, Sandwell, Solihull, Walsall, Wolverhampton*

Brunel House
2 Fitzalan Road
Cardiff
CF2 1SH

Tel: 01222 473777

Covers: *Clwyd, Dyfed, Gwent, Gwynedd, Mid Glamorgan, Powys, South Glamorgan, West Glamorgan*

The Marches House
Midway
Newcastle-under-Lyme
ST5 1DT

Tel: 01782 717181

Covers: *Hereford & Worcester, Shropshire, Staffordshire*

Birkbeck House
Trinity Square
Nottingham
NG1 4AU

Tel: 0115 9470712

Covers: *Derbyshire, Lincolnshire, Nottinghamshire*

Sovereign House
110 Queen Street
Sheffield
S1 2ES

Tel: 0114 2739081

Covers: *Barnsley, Doncaster, Humberside, Rotherham, Sheffield*

8 St Pauls Street
Leeds
LS1 2LE

Tel: 0113 2446191

Covers: *Bradford, Calderdale,
Kirklees, Leeds, Wakefield, Craven,
Hambleton, Harrogate,
Richmondshire, Ryedale,
Scarborough, Selby, York*

Quay House
Quay Street
Manchester
M3 3JB

Tel: 0161 831 7111

Covers: *Bolton, Bury, City of
Manchester, City of Salford, Oldham,
Rochdale, Stockport, Tameside,
Trafford, Wigan*

The Triad
Standley Road
Bootle
Merseyside
L20 3PG

Tel: 0151 922 7211

Covers: *Chester, Congleton, Crewe,
Ellesmere Port, Halton, Knowsley,
Liverpool, Macclesfield, St Helens,
Sefton, Vale Royal, Warrington, Wirral*

Victoria House
Ormskirk Road
Preston
PR1 1HH

Tel: 01772 259321

Covers: *Cumbria, Lancashire*

Arden House
Regent Centre
Regent Farm Road
Gosforth
Newcastle-upon-Tyne
NE3 3JN

Tel: 0191 284 8448

Covers: *Cleveland, Durham,
Newcastle-upon-Tyne,
Northumberland, North Tyneside,
South Tyneside, Sunderland*

Belford House
59 Belford Road
Edinburgh
EH4 3UE

Tel: 0131 247 2000

Covers: *Borders, Central, Fife,
Grampian, Highland, Lothian,
Tayside, Orkney and Shetland*

375 West George Street
Glasgow
G2 4LW

Tel: 0131 275 3000

Covers: *Dumfries and Galloway,
Strathclyde and the Western Isle*

Appendix 3

Model risk assessment form

Management of Health and Safety at Work Regulations

RISK ASSESSMENT RECORD

Play Project:

Date of Assessment:

Work Activity to be Assessed:

Name of Assessor:

Position:

1. Activity/ Process Occupation	2. What hazards to health and/or safety exist?	3. What risks do they pose to Employees and other Persons?	4. Precautions already taken?	5. Risk Level achieved? (High, Medium or Low)	6. Are additional measures necessary? NOTE: this Section must be completed if Risk Level is HIGH.

Are any Special Groups at risk? YES/NO If YES, who are these and how many?

©PLAYLINK

Appendix 4

Useful contacts

Child Accident Prevention Trust
4th Floor
Clerks Court
18-20 Farringdon Lane
London
EC1R 3AU

(0171) 608 3828

Children's Legal Centre
The University of Essex
Withenhoe Park
Colchester
Essex
CO2 3SG

(01206) 873 820

Children's Play Council (formerly
NVCCP)
c/o NCB
8 Wakely Street
London
EC1V 7QE

(0171) 843 6000

Community Transport Association
Highbank
Halton Street
Hyde
Cheshire
SK14 2NY

(0161) 351 1475

Council for Outdoor Education
Training and Recreation
The Muncaster Country Guest House
Ravenglass
Cumbria
CA18 1RD

(01229) 717693

Council for Outdoor Education
Training and Recreation Scottish
Council
Lagganlia Centre for Outdoor
Education
Kincraig Kingussie
Invernesshire
PH21 1NG

(01540) 651265

HAPA – Adventure Play for Children
with Disabilities and Special Needs
Pryor's Bank
Bishops Park
London
SW6 3LA

(0171) 736 4443

Health and Safety Executive
Rose Court
2 Southwark Bridge
London
SE1 9HS

(0171) 717 6000

HSE (Health and Safety Executive)
Books
PO Box 1999
Sudbury
Suffolk
CO10 6FS

(01787) 881165 / Public Enquiry
Point (0114) 284 2345

HMSO
Central Orderings – (0171) 873 9090

Institute of Leisure and Amenity
Management
ILAM House
Lower Basildon
Reading
Berkshire
RG8 9NE

(01491) 874222

Institute of Materials
1 Carlton House Terrace
London
SW1Y 5DB

(0171) 839 4071

Kid's Clubs Network
Bellerive House
3 Muirfield Crescent
London
E14 9SZ

(0171) 512 2100

National Children's Bureau
8 Wakely Street
London
EC1V 7QE

(0171) 843 6000

National Council for Voluntary
Organisations
26 Bedford Square
London
WC1B 3HU

(0171) 713 6161

National Play Information Centre
199 Knightsbridge
London
SW7 1DE

(0171) 584 6464

National Playing Fields Association
25 Ovington Square
London
SW3 1LQ

(0171) 584 6445

PLAYLINK
The Co-op Centre
Unit 5 Upper
11 Mowll Street
London
SW9 6BG

(0171) 820 3800

Royal Society for the Prevention of
Accidents
Edgbaston Park
353 Bristol Road
Birmingham
B5 7ST

(0121) 248 2000

Appendix 5

Closing down checklist

Tasks to be undertaken at closing time:

- Ensure that all children are off site
- Immobilize or dismantle any moving equipment
- Ensure that stacked materials do not represent a hazard
- Remove ladders and other items of loose equipment to a secure store
- Remove all tools and play equipment to storage
- Empty waste containers and lock bins
- Board over any holes or excavations
- Secure animal cages/areas
- Cover sand areas
- Empty pools
- Extinguish fires
- Lock all buildings or stores and outer gates
- Turn off and unplug electrical appliances (except those designed to be left on, check with the manufacturer)
- Turn off services at the mains where practicable
- Arm alarms
- If closure is exceptional, post notices.

Appendix 6

Site checklist

Effective safety measures are dependent upon good maintenance procedures. Regular inspections are essential to identify problems, potential hazards and maintenance requirements.

The following checklist only provides a basic framework and requires adaptation to the requirements of each individual site.

It is recommended that at least one full independent inspection per year is undertaken by a person/persons competent to do so and that a full report on the findings be made to the operators of the site. PLAYLINK is able to undertake these inspections.

ITEM	DAILY	WEEKLY	MONTHLY/ OTHER
FENCING			
Is fencing complete and secure?	Visual check	–	
Are posts/stanchions solid?	–	Thorough physical check	
Is fencing free of protrusions, etc?	Visual check	Thorough inspection	
GATES			
Are gates in good condition?	Visual check	–	
Do gates open and close easily?	Visual check and use	Vehicle accesses and other gates not used daily should be checked in use at least weekly.	
Are linings in good order?	Visual check and use	–	
NOTICES			
Are notices up to date and readable?	Visual check	–	
ACCESS			
Is pedestrian access free of obstruction?	Visual check	–	
Is vehicle access free of obstruction?	Visual check	–	
Are surfaces in good condition?	Visual check	–	
Is there access for people with disabilities?	Visual check	–	

ITEM	DAILY	WEEKLY	MONTHLY/OTHER
PLAYGROUND SITE			
Is site generally free of debris and hazardous rubbish?	Close visual check	–	
Is site free of litter?	Visual check	–	
Is site free of surface water?	Visual check	–	
SURFACING			
Are pathways in good condition?	Visual check	–	
Are general play surfaces in good condition?	Visual check	–	
Are impact absorbing surfaces clean and free of debris (solid and loose fill)?	Visual check (solid surfaces) and raking through (loose fill surfaces)	Rake and turn loose fill IAS	
Are impact absorbing surfaces sufficiently deep (loose fill)?	Check by measuring	–	Check by measuring
STORAGE			
Are boundaries to storage areas secure?	Visual check	–	
Is wood stored away from contact with the ground?	Visual check	–	
Are materials stacked safely?	Visual check	–	
Is material available for use in a safe condition?	Visual check	–	
Are storage areas for hazardous material signed and secure?	Check by testing	–	

© PLAYLINK

ITEM	DAILY	WEEKLY	MONTHLY/ OTHER
REFUSE Are waste receptacles in position?	Visual check	–	
Are waste receptacles empty and ready for use?	Adequate arrangements for removing rubbish must be made – frequency of removal must depend on individual site factors	Adequate arrangements for removing rubbish must be made – frequency of removal must depend on individual site factors	Adequate arrangements for removing rubbish must be made – frequency of removal must depend on individual site factors
Are areas around waste receptacles free of rubbish?	Visual check	–	
FIRE AREA/BBQ Is fire area free of excess ash?	–	Visual check remove as necessary	
Is fire area free of combustible debris?	Visual check	–	
Is firewood free from chemicals such as timber preservatives and neatly stored a safe distance from the fire?	Visual check	–	
Are adequate means of extinguishing the fire available by the fire area?	Visual check. Fire areas require very high levels of supervision when a fire is burning. A constant check must be kept to ensure that means to extinguish the fire rapidly are available.	–	

ITEM	DAILY	WEEKLY	MONTHLY/OTHER
SAND AREA/SAND BOX Are sand areas clean and free of debris and contamination?	Visual check and by thorough raking?	Disinfect thoroughly	Turn over to full depth
Is the sand sufficiently deep?	–	–	Test by measuring
Is the sand area cover adequate?	Visual check	–	
WATER Are water areas clean and free of debris?	Visual check		
Is the area surrounding water areas clean and free of debris?	Visual check		
Is the water clean?	Visual check		
Is the circulation/filtration installation working adequately?	Check by testing in accordance with manufacturers' instruction		
Does the water meet hygiene standards?			Initial check with Environmental Health Officer and follow-up as required
HOLES AND EXCAVATIONS Are all holes to be re-used safely covered?	Visual check and test for solidity		
Are holes free of refuse?	Visual check		
Is appropriate shoring in place?	Visual check		

ITEM	DAILY	WEEKLY	MONTHLY/OTHER
STRUCTURES			
Are all uprights stable and in good condition?	Visual check	Stress to test for excessive movement	It is recommended that a separate schedule of checks be prepared for each structure or distinct parts thereof. The checks listed apply to all types of structure but to avoid repetition features relating to particular types (e.g. slides) are listed. Checks should include the surface around the equipment and the condition of structures to a depth of 300 mm below ground.
Are all bracing members in good condition?	Visual check	Stress to test for excessive movement	
Are all horizontal structural members in good condition?	Visual check	Test for rigidity and adequate bracing	
Are all joists in good condition?	Visual check		
For new equipment, have post-installation and running-in inspections taken place?			
Is all planking and floor in good condition and firmly fixed?	Visual check and test by use	Close check for damage	Close check for damage and secure fixing
Are all handrails in good condition?	Visual check and test by use	Stress to test for excessive movement	

ITEM	DAILY	WEEKLY	MONTHLY/ OTHER
STRUCTURES continued Are all kick/toe boards in good condition?	Visual check and test by use		
Is all boarding, paling and solid infilling in good condition?	Visual check	Close check for damage and secure fixing	
Are all nailed joints secure?	Visual check and test by use	Stress to check for movement	
Are all screwed joints secure?	Visual check	Stress to check for movement	
Are all bolted joints secure?	Visual check	Check all nuts for tightness. Stress to check for movement	Re-tighten all nuts. Check for damage and signs of lateral stressing of bolts.
ACCESS Are ladders securely fixed?	Visual check and test by use	Stress to check for movement	
Are all ladder rungs intact and securely fixed?	Visual check and test by use	Stress to check for movement	
Are all stairways securely fixed?	Visual check and test by use	Stress to check for movement	
Are all steps intact and securely fixed?	Visual check and test by use	Stress to check for movement	

ITEM	DAILY	WEEKLY	MONTHLY/ OTHER
ACCESS continued			
Are all nets securely fixed?	Visual check and test by use	Stress to check for movement	
Are all nets in good condition?	Visual check and test by use		
Are all sliding poles securely fixed at top and bottom?	Visual check and test by use	Stress to check for movement	
Are sliding poles free of rust, surface blemishes or damage?	Close visual check then check by use		Smooth surface appropriately
Are ropes securely fastened?	Visual check and test by stressing		
Are ropes free from wear by friction?	Visual check		
Are ropes in good condition, free of unwanted knots, fraying at ends and other damage?	Close visual inspection	Remove ropes to undertake thorough close examination	

ITEM	DAILY	WEEKLY	MONTHLY/ OTHER
SWINGS MAIN STRUCTURE			
Are all swings securely fixed?	Close visual check when in use		Very close examination of all elements
Are all friction points protected?	Visual check for wear		Lubricate as necessary, according to levels of use
Are all loops/eyes protected by thimbles?	Visual check		
Are all shackles in good condition and securely closed?	Close visual check for wear and test for tightness		
Are all chains in good condition and free of excess wear on links?	Visual check	Close check of individual links	
Are all chain fixings secure?	Visual check and check for tightness		
Are all wire ropes in good condition?	Visual check	Check for damage (wear gloves)	
Are wire rope fixings secure?	Visual check and test for tightness		
Are ropes in good condition, free of unwanted knots, fraying at ends, and other damage?	Close visual inspection	Remove ropes to undertake thorough visual inspection	
Are swing seats in good condition and securely fixed?	Visual check and test by use		

ITEM	DAILY	WEEKLY	MONTHLY/OTHER
SLIDES			
Are supports for slide solid and stable?	Visual inspection and test for movement	Stress to check for movement	
Is slide surface free of damage, splinters, protrusions, etc.?	Very close inspection when in use	Clean and smooth surface appropriately	
Are sides of slide free of damage, splinters, protrusions, etc.?	Very close inspection	Clean and smooth appropriately	
Are sides of slide adequately supported?	Visual inspection and test for movement	Stress to check for movement	
AERIAL RUNWAYS			
Are runway anchorages stable and solid?	Visual check (while runway is in use) to check for movement	Stress to check for movement	
Is the supporting structure for the running line solid and securely fixed?	Visual check (while runway is in use) to check for movement	Stress to check for movement	
Are all points of contact suitably protected?	Visual check for evidence of wear		
Is running line securely fastened at each anchor point?	Visual check and check for tightness		
Is running line at suitable angle to control rate of descent?	Test by use, adjust as necessary		

ITEM	DAILY	WEEKLY	MONTHLY/ OTHER
AERIAL RUNWAYS continued			
Is running line in good condition free of damage and wear?	Close visual check	Very close examination and greasing as required (NB wear gloves)	The whole running line should be dropped at least once a year for thorough examination. Lubricate as necessary. (NB wear gloves)
Is pulley block in good condition?	Visual check		
Is pulley block securely locked in position when in use?	Test by stressing		
Is pulley wheel in good condition?	Visual check		
Is pulley wheel spindle free of wear?		Check by dismantling	
Is handgrip in good condition?	Visual check and test by use	Check by dismantling	

Appendix 7 Building checklist

ITEM	DAILY	WEEKLY	MONTHLY/ OTHER
HEATING			
Are all heating appliances in good working order?	Test by use when in use		
Are all guards securely fixed and in good condition?	Visual check		
Are heating levels sufficient?	Temperature check where applicable		
Are thermostats fitted in a secure position?	Visual check		
ELECTRICAL SYSTEM			
Are all switches and sockets working?	Test by use as required		
Are sockets protected with lockable covers?	Visual check		
Has annual check been undertaken by IEE registered electrician?			
Is low voltage equipment in use?	Visual check		
Is the circuit protected by an ELB or RCD?	Test by use as required		
Has portable equipment been checked?	See Appendix 8		See Appendix 8

ITEM	DAILY	WEEKLY	MONTHLY/OTHER
GAS			
Are all piliot lights lit?	Test before lighting flame		
Are all gas applicances working?	Test by use		
Is there any evidence of gas leakage?	Test by smell		
WATER			
Is water at the right temperature?	Test by use as required		
Is there evidence of water leakage?	Visual check and test by use		
Are all taps in good working order?	Visual check and test by use		
Are all cisterns and tanks in good condition and working?	Test by use as required		
Are waste pipes clear of obstruction?	Visual check		
STAIRWAYS			
Are all stairways secure and in good condition?	Stress to test for movement		
Are handrails and infill panels securely fixed and in good condition?	Visual check and stress to test for movement		
Are stairways clear of obstruction?	Visual check		
Are stair treads free of debris, spillage, etc?	Visual check		

ITEM	DAILY	WEEKLY	MONTHLY/OTHER
PASSAGEWAYS Are passageways clear of obstruction?	Visual check		
STRUCTURAL Is there evidence of damage by damp or rot?			Visual check
TOILETS Are all toilets thoroughly cleaned?	Check at least once a day		
Are all cisterns, WCs, urinals and basins clean and in good working order?	Check and test		
Are hand-drying facilities working?	Visual check		
Are there adequate supplies of toilet paper?	Replenish as necessary		
Are waste receptacles cleared and clean?	Empty as necessary		
Are floors clean and dry?	Visual check		
Are all waste and drainage outlets clear?	Test by use		
STORAGE AREAS Are storage facilities for hazardous materials and equipment secure and properly signed?	Visual check		
Are materials and equipment stored safely?	Visual check		
Is access to stored materials and equipment free of obstruction?	Visual check		

ITEM	DAILY	WEEKLY	MONTHLY/ OTHER
KITCHENS Are requirements of food safety legislation met?	Check refrigerator/freezer temperatures		Initial expert check and follow-up required
Are floors and surfaces clean and free of spillage?	Visual check		
Are all utensils and equipment clean?	Visual check before use		
Are cooking facilities clean and free of spillage and vermin?	Visual check		
Is fire-fighting equipment in good condition and accessible?	Visual check		See section on fire precautions
Are basins or sinks clean?	Visual check		
Are taps in good working order?	Test by use		
Are waste outlets clear?	Test by use		
SPECIAL ACTIVITY AREAS Are floors and surfaces clean and free of spillage?	Visual check		
Is equipment safely stored and labelled?	Visual check		
Are materials safely stored?	Visual check		
Is all equipment in good condition?	Test before use		

ITEM	DAILY	WEEKLY	MONTHLY/OTHER
FIRST AID FACILITIES Is the area set aside for treatment clean?	Visual check		
Are stocks of first aid equipment adequate?	Detailed visual check of stock levels against list of requirements	Thorough inventory	
FIRE PRECAUTIONS Are fire procedures and signs displayed?	Visual check		All fire extinguishers must be maintained and tested by a competent trained person in accordance with manufacturer's instructions and at least once a year
Are all items of fire-fighting equipment in position and ready for use?	Visual check		
Has a fire drill(s) been carried out?			Fire drills at least twice a year
Are fire exits unlocked and exits unlocked?	Visual check		
Are smoke detectors fitted, clean and working?		Audible test	Visual test and clean once a month, new batteries once a year

© PLAYLINK

ITEM	DAILY	WEEKLY	MONTHLY/ OTHER
GENERAL Are floors clean and free of spillage?	Visual check		
Are floor surfaces in good condition?	Visual check		
Are walls clean and in good condition?	Visual check		
Are surfaces clean and in good condition?	Visual check		
Is furniture in good condition and fire safe?	Visual check and test by use		Check labels on purchase or arrival on site
Are waste receptacles cleared and clean?	Visual check		
ROOFS Is there any external damage?			Visual checks
			External damage may be disclosed only by internal signs such as evidence of leaking
DRAINAGE Are gutters securely fixed and undamaged?	Visual check		Visual check

ITEM	DAILY	WEEKLY	MONTHLY/OTHER
DRAINAGE continued Are downpipes securely fixed and undamaged?	Visual check		Gutters will require more attention in the autumn when they may be blocked by leaves, and when rainfall is heavy
Are drainage points clear of debris or other blockage?	Visual check		
Are gutters clear of debris or other blockage?	Visual check		
WALLS Is there any external damage?			Visual check
Is there any evidence of damp?			Visual check
DOORS Are external doors securely fixed and in good working condition?	Test by use		
Are internal doors in good working condition and do they meet fire regulations, including door closers working	Test by use		
Are locks and door fixings secure and in good working order?	Test by use		
Are all glazed door panels secure and intact, and fitted with safety glass where required?	Visual check		

ITEM	DAILY	WEEKLY	MONTHLY/OTHER
LIGHTING			
Are all lights working, and diffusers in place?	Test by use		Fluorescent tubes and diffusers should be cleaned periodically to maintain illumination levels
Are emergency lighting systems working?	Test by use		
WINDOWS			
Are all windows securely fixed?	Test by use as required		
Is all glazing material securely fixed and intact?	Visual check		
Do windows open and close easily without endangering operator or users?	Test in use as required		
Are all fittings secure and in good working order?	Test in use as required		
Are guards or shutters securely fixed, easy to operate and in good condition?	Visual check		
Are windowsills clear?	Visual check		

Appendix 8

Routine checks for portable electrical equipment

Listed below are typical routine electrical checks for portable apparatus, to be carried out by a suitably competent person, reproduced from HSE Guidance Note PM 32.

Note: This checklist is intended as a guide to what may be required; certain apparatus may need different or additional inspections and tests. Non-electrical checks are outside the scope of this guidance note.

Equipment:
Make:
Serial No:

Item	Test	Pass condition
1 mains lead	a) visual inspection	two layers of insulation no damage
	(b) mains plug	correctly connected cable clamp gripped to sheath correct fuse fitted
2 either: mains lead or instrument connector (if lead detachable)	a) visual inspection of instrument male connector	IEC 320 type or equivalent (BS4491, CEE22),
	(b) attempt to open socket without tool	unopenable
	(c) attempt to pull socket cable from female connector	no movement
	(d) polarity of 3-pin units	as per BS 4491
or: grommet/clamp	(a) inspection of grommet	cable insulation protected

Item	Test	Pass conditions
	(b) sharp pull on cable	no appreciable movement
	(c) rotation of cable	no rotation
3 mains on/off switch	(a) visual inspection no damage	no damage correct operation
either 4 and 5 or 6 and 7	(a) visual inspection:	
4 conducting case	(if marked ⊗ treat as item 5)	
	Earth tester which will check resistance and pass a current of at least twice the fuse rating	earth resistance 0.1 Ω or ohms; or earth resistance 0.5Ω or ohms for loads fused at 3 amps or less
	b) high voltage insulation 500 V ac minimum test	no fault indicated after 5 seconds
5 insulation case	visual inspection	maker's double insulation mark visible case undamaged – if in doubt test using portable Appliance Tester
6 accessible fuse holders	visual inspection	no damage, removal of carrier does not permit live part to be touched
7 exposed output connections	(a) visual inspection	no voltage greater than 50 V
	(b) for outputs greater than 50 V˙	short-circuit less than 5 mA or short-circuit current greater than 5 mA and labelled 'unsuitable for use by children'

˙i.e. live at more than 50 V when in use.

NOTE: at least 25% of all double insulated equipment should be tested each year, i.e. all equipment is tested at least once every 4 years.
Overall result (delete as necessary)

Failed

Unit is   ....

Passed

Signed   ....   Date   ....

This appendix may be freely reproduced in part or in full, other than for the purposes of commercial reproduction, provided the source is acknowledged.

Reproduced with the permission of the Health and Safety Executive.

Appendix 9

Incubation and Exclusion Periods of the Common Infectious Diseases

Disease	Usual incubation period (days)	Interval between onset of illness and appearance of rash (days)	Minimum period of exclusion provided child appears well	
			Patients	Family contacts
Chicken pox	10–21	0–2	Seven days from appearance of rash; all the scabs need not have separated	There is no routine exclusion of contacts of any of these infectious diseases but individual children may be excluded on the advice of a General Medical Practitioner.
Dysentery	1–7	–	Until 24 hours after cessation of diarrhoea	
Food poisoning	0–2	–	Until declared fit	
German measles	14–21	0–2	Until clinical recovery	
Infective jaundice	14–42	–	Until clinical recovery	
Measles	7–21	3–5	Until clinical recovery	
Meningitis (bacterial) (viral)	2–10 0–21	– –	Until clinical recovery and bacteriological examination is clear	
Mumps	12–28	–	Until disappearance of all swelling	
Scarlet fever	2–5	1–2	Until clinical recovery	
Whooping cough	5–14	–	Until clinical recovery	

Remember to warn all parents in the group if a case of German measles is confirmed. It is most dangerous in the very early stages of pregnancy

EXCLUSION PERIODS OF THE COMMON INFECTIONS

Disease	Minimum period of exclusion
Impetigo	Until spots have healed, unless lesions can be covered.
Pediculosis (head lice)	Until treatment has been carried out.
Verrucae (plantar warts)	Exclusion from barefoot activities until certified free from infection
Ringworm of feet (Athlete's foot)	Exclusion from barefoot activities until certified free from infection.
Ringworm of scalp or body	Until adequate treatment instituted, provided lesions are covered.
Threadworm	Until adequate treatment instituted.
Scabies	Until adequate treatment instituted.

A General Medical Practitioner or Local Health Clinic will be able to give further advice.

Reproduced from *Guidelines for Good Practice for Sessional Playgroups*, Pre–School Playgroups Association – with their kind permission.

Appendix 10

Contents of first-aid boxes

Item	Number of users		
	11–50	51–100	101–150
Guidance card	1	1	1
Individually wrapped sterile adhesive dressings	40	40	40
Sterile eye pads, with attachment	4	6	8
Triangular bandages	4	6	8
Sterile coverings for serious wounds	4	6	8
Safety pins	12	12	12
Medium sized sterile unmedicated dressings	8	10	12
Large sterile unmedicated dressings	4	6	10
Extra large sterile unmedicated dressings	4	6	8

Where sterile water or sterile normal saline in a disposable container needs to be kept near the First Aid Box because tap water is not available, at least the following quantities should be kept:

Sterile water, or saline in disposable 300 ml containers	3	6	6

Soap, water and disposable drying materials should also be available. In the absence of soap and water, other cleaning methods, such as sterile wipes, may be used.

Disposable plastic gloves, aprons and other suitable protective equipment should be provided near the first-aid materials and should be properly stored and checked regularly to ensure that they remain in good condition.

Plastic disposal bags for soiled or used first-aid dressing should be provided.

Blunt-ended, stainless steel scissors (minimum length 12.70 cm) should be kept with the protective clothing and equipment.

Appendix 11

Travelling first-aid kits

The contents of travelling first-aid kits should be appropriate for the circumstances in which they are to be used. At least the following should be included:

- **Card giving the general first-aid guidance**
- **Six individually wrapped, sterile, adhesive dressings**
- **One large sterile unmedicated dressing**
- **Two triangular bandages**
- **Two safety pins**
- **Individually wrapped, moist, cleansing wipes**

Procedures for prevention of contamination by blood/body fluids

Skin
Cuts or abrasions in any area of exposed skin should be covered with a dressing that is waterproof, breathes and is an effective viral and bacterial barrier.

Gloves
Non-sterile, seamless, latex or vinyl gloves should be worn during non-invasive procedures where there may be contamination of hands by blood/body fluids.

Hand washing
The use of gloves does not preclude the need for thorough hand-washing between procedures.

Aprons
Disposable plastic aprons may be worn if there is a possibility of splashing by blood/body fluids.

Eyes
Where there is a danger of flying contaminated debris or blood splashes, eye protection is necessary.

Sharps
Extreme care must be exercised during the disposal of sharps which should be disposed of into approved sharps boxes. These should never be over-filled.

Needle-stick injury
Concentrate on the procedure to avoid this happening. In the event of a sharps or needle-stick injury:

- Encourage bleeding from the puncture wound. Do not suck.
- Wash the area thoroughly with soap and water.
- Cover with waterproof dressing.
- Report accident/notify line manager.

Spillages of body fluids (ie blood, urine, vomit) on floors
Wear disposable latex gloves and a plastic apron. Using Presept granules, sprinkle over the spilled body fluid sufficient to solidify and disinfect the spillage. Cover with disposable towels and leave for 30 minutes. The spillage should be cleared up with the gloved hand and debris treated as clinical waste. The area should then be cleared with the appropriate domestic cleaning product for that surface.

Waste
All contaminated waste must be placed into yellow clinical sacks and left for incineration.

The yellow biohazard bag must be collected by the Environmental Health Department for incineration.

Cleaning of surfaces
Dissolve 1 Presept tablet to 1 litre of water. Wearing disposable vinyl gloves, wipe down contaminated areas with paper towels soaked in Presept. Paper towels, gloves, etc must be disposed of in a yellow biohazard bag, which must be collected by the Environmental Health Department for incineration.

REFERENCES

DoH (1990) *Guidance for Clinical Health Care Workers, Protection Against Infection with HIV and Hepatitis Viruses: Recommendation of the Expert Advisory Group on AIDS*, London HMSO.
Information published by the Royal College of Nursing 20 Cavendish Square London W1M OAB, update June 1993.

Appendix 13

Accident Report Form

The following accident report form should be completed in triplicate immediately following treatment for any injury at the adventure playground. One copy must be held in the records of the playground, one copy should be made available to the parent(s) or guardian(s) of the injured/treated person (if appropriate) and a copy should accompany the injured person if they are referred for further treatment.

Adventure playground . . . .

Full address . . . .

Tel. no: . . . .

Name (injured person) . . . .

Address . . . .

. . . .

Tel. no: . . . .

Age . . . .

Nature of injury . . . .

. . . .

. . . .

. . . .

Treatment given . . . .

. . . .

. . . .

Further action taken :   none
(delete where not applicable) : Parent(s)/carer(s) informed
                                 Taken home
                                 Referred to doctor (specify below)
                                 Referred to hospital (specify below)
                                 Further treatment advised (specify below)

Further action. . . . .

. . . .

. . . .

Injured person's GP (Name and address) . . . .

. . . .

. . . .

Any known allergy. . . . .

Any known medical condition . . . .

Any known medication. . . . .

┌───┐
│ FOR PLAYGROUND RECORDS ONLY   │
│   │
│ Probable cause of injury . . . . │
│   │
│ Witnesses . . . . │
│   │
│ Remedial action taken . . . . │
└───┘

©PLAYLINK

Model off-site activity consent form

NB. THIS CONSENT FORM IS IN TWO PARTS:
PART ONE to be retained by parent or guardian
PART TWO to be returned to the playground worker in charge

PART ONE
Name of organization . . . .

Name of playground . . . .

Telephone number of playground . . . .

Activity . . . .

Date . . . .

Site leader . . . .

Details of the off-site activity

To . . . .

Date(s) . . . .

Departure time: . . . .
From . . . .

Return time: . . . .
To . . . .

Mode of transport ....

Cost per child ....

Party leader ....

Ratio of children to supervisors ....

No more than .... children to.................... supervisors

Closing date for return of form ....

Among the activities will be (full details of planned/likely/possible activities):

....

....

....

....

* Further details can be obtained from the playworker in charge
or
* There will be a full briefing for parents on

Date ....

Time ....

Place ....

Please attend.
Delete whichever is not applicable.

Should you need to contact your child/children in an emergency when the centre is closed telephone ....

Please note: While this organization has insurance provision in respect of any accident to children while taking part in this off-site activity, which is caused by the negligence of its servants or agents, they do not make any insurance provision for personal accidents to children where no negligence is involved. Parents/carers may wish to consult their own insurers to arrange insurance for personal accidents.

PART TWO
Please complete the sections below and return this part of the form to the playworker in charge. Keep Part One of the form for your information.

Activity . . . .

Date . . . .

Full name of child . . . .

Date of birth . . . .

. . . .
may attend the above activity. I understand that while he/she will be under the control of the playworker and/or other supervisors approved by the organization and that, while the staff in charge of the party will take all reasonable care of the children, they cannot necessarily be held responsible for any loss, damage or injury suffered by my child arising during, or out of, this journey.

I agree to my child taking part in any, or all, of the activities described above.

I consent to any emergency treatment that may be necessary during the course of the off-site activity, provided that, in the opinion of the Doctor, any delay required to obtain my signature may endanger my child's health and safety.

My child has/has not* been actively sensitive to penicillin.

My child suffers from . . . .

. . . .
requiring regular treatment.

My child's home doctor is . . . .

Phone number . . . .

Emergency number . . . .

The Party Leader should be aware of the following medication/medical condition(s) of my child which may impair his/her ability to take part in this activity.

Condition . . . .

Proposed activities which may be affected . . . .

. . . .

. . . .

. . . .

My child is/is not* able to swim 25 metres.

*Delete whichever is not applicable.

Name of parents/carer (BLOCK CAPITALS PLEASE) . . . .

. . . .

Address . . . .

. . . .

. . . .

. . . .

Telephone number:

daytime . . . .

evenings . . . .

Signature of parent/carer . . . .

Date . . . .

Appendix 15

Minibus drivers' and escorts' checklist

WHAT SHOULD A DRIVER/ESCORT DO? – COMMUNITY TRANSPORT ASSOCIATION ADVICE

- Make sure that children are supervised when boarding or leaving the minibus; take particular care if children have to exit from the back of the minibus. Plan which passengers will sit in the front seats and by the doors.
- Never allow passengers to board/leave until the vehicle is at a complete standstill, and safely parked by an adjacent pavement or other traffic-free area.
- Do not drive away until everyone is seated comfortably, facing the front and wearing a seat-belt (when fitted). If in special seats, or using booster cushions ensure these are properly fixed and adjusted.
- Escorts should sit where they can see the passengers – not in the seat adjacent to the driver.
- Always ensure that ambulant disabled passengers are seated safely and comfortably and that anyone travelling in a wheelchair is also safely secured. Wheelchairs not in use must be securely stored.
- Make sure that you have a complete list of the passengers being carried with a note of any special medical or other needs.
- Keep such a list with other relevant documents in a place where it can be readily found in the event of an accident.
- Take care when using passenger lifts and other specialist equipment. Always comply with the manufacturer's instructions.
- Do not allow boisterous play of any kind.
- Always park so that children or elderly passengers can step onto the footway, not onto the carriageway.
- Take particular care when reversing the vehicle if children are nearby. (Nearly a quarter of all deaths involving vehicles at work occur while the vehicle is reversing.) Avoid reversing whenever possible. If it is unavoidable, seek adult

assistance to exclude children from the area, move slowly and keep the reversing distance to a minimum.

- Do not allow the children to operate doors.
- Approach each stop slowly and with care.
- Flashing lights, if fitted, should only be used when passengers are boarding or alighting.
- When 'school bus' signs are used, make sure they are in position while pupils are being transported, and that they do not obstruct the driver's vision.
- Check that no bags or clothing are caught in the doors.
- Check all the mirrors every time before moving away, in case latecomers are approaching the vehicle.
- Check that no luggage is unsecured, or could pose a danger.
- If the vehicle breaks down, or in the event of an accident, give clear instructions to the passengers and see that children remain supervised: their safety is paramount. If there is a risk of fire, however small, evacuate the vehicle, moving the occupants to a safe place. On motorways, never stop anywhere on the carriageway; use the hard shoulder.
- If there is any serious delay during the journey, inform the school/organization so that information may be passed to parents.
- Children should not be allowed to push vehicles for any reason; this could result in accident or injury.
- Enforce a 'no smoking' rule.
- All children under five should travel with a driver and/or escort known to them.
- Children must not be left unaccompanied in the minibus.
- Supervision should include noise control – keep the noise levels down.
- Never use or threaten a child (or adult) with physical force. If behaviour is unsuitable, stop the vehicle when it is safe to do so and get assistance. If you threaten to report problem behaviour, then do so.
- Ensure that adequate arrangements are made for toileting before and during the journey.
- Plan breaks, and avoid long spells of driving when children might get restless and bored.
- Try to keep children occupied – the journey will seem much quicker.
- Make provision for travel sickness, with pills (where parents provide and request), bags or buckets.
- Ensure that litter is disposed of carefully (provide a litter bag or bin – cans rolling around the floor can be extremely distracting).

Reproduced with the kind permission of the Community Transport Association from *The Operation of Minibuses in the Voluntary Sector*.

Appendix 16

Travelling by minibus: some rules for children

- Be punctual, and wait for the minibus away from the road in a sensible manner.
- Do not push, or rush towards the vehicle when it arrives, as you may push someone into the path of the moving vehicle and the driver may be unable to stop.
- Find a seat quickly and quietly without pushing.
- Always remain seated when the vehicle is moving and wear your seatbelt whenever they are fitted.
- Do not kneel on your seat. This is dangerous and accidents have resulted from it.
- Only speak to the driver when he or she is not driving, or in an emergency, and remember that too much noise can disturb the driver and cause an accident.
- There must be no throwing of things, or any other kind of horseplay.
- Wait until the vehicle has stopped before getting up to leave.
- Make sure your belongings do not obstruct the gangway or use up valuable seat space.
- Make sure you have your belongings with you when you leave the vehicle, but be prepared to leave your belongings behind in an emergency.
- If you have forgotten something, take care if you return to the vehicle, as the driver may be pulling away. In that case, tell the supervisor what has happened.
- If you have to cross the road after getting off the minibus, always allow the minibus to move off before attempting to cross, and always use the Green Cross Code.
- Go to the toilet before you get on the bus. If you need to use the toilet, or are feeling unwell, don't be afraid to say so to the supervisor as early as possible.

Reproduced with the kind permission of the Community Transport Association from *The Operation of Minibuses in the Voluntary Sector*.

Appendix 17

Passenger lifts/ramps – Community Transport Association advice

- Code of Practice VSE87/1 (Department of Transport) (see Section 9 – Useful Sources of Information) sets out the relevant requirements for power-operated lifts and also for ramps. These require that instructions on how to use the lift be displayed and the controls clearly marked and accessible from both inside and outside the vehicle.
- Make sure brakes are on.
- Make sure the passenger knows what you are about to do.
- Electric wheelchairs should be in manual mode and full assistance given to passengers using such chairs, especially when the chair has to be reversed off the vehicle.
- Weight limits should be adequate to take even the heaviest chair, passengers and helper. Watch out for feet!
- Ramp gradients should not exceed 1 in 12, but this is not always practicable.
- Non-slip material should be used on the ramps and raised edges should be provided.
- All ramps must be provided with a secure means of stowage when not in use, but should not be stowed in a way that would prevent the opening of doors in an emergency.

Reproduced with the kind permission of the Community Transport Association from *The Operation of Minibuses in the Voluntary Sector*.

Appendix 18

Good handling technique

The development of good handling technique is no substitute for other risk reduction steps such as improvements to the task, load or working environment, but it will form a very valuable adjunct to them. It requires both training and practice. The training should be carried out in conditions that are as realistic as possible, thereby emphasizing its relevance to everyday handling operations.

The content of training in good handling technique should be tailored to the particular handling operations likely to be undertaken. It should begin with relatively simple examples and progress to more specialized handling operations as appropriate. The following list illustrates some important points, using a basic lifting operation by way of example.

Stop and think
- Plan the lift. Where is the load going to be placed?
- Use appropriate handling aids if possible.
- Do you need help with the load?
- Remove obstructions, such as discarded wrapping materials.
- For a long lift – such as floor to shoulder height – consider resting the load mid-way on a table or bench, in order to change grip.

Place the feet
- Feet apart, giving a balanced and stable base for lifting (tight skirts and unsuitable footwear make this difficult). Leading leg as far forward as is comfortable.

Adopt a good posture
- Bend the knees, so that the hands when grasping the load, are as nearly level with the waist as possible.
- Do not kneel or overflex the knees.
- Keep the back straight (tucking in the chin helps).
- Lean forward a little over the load if necessary to get a good grip. Keep shoulders level and facing in the same direction as the hips.

Get a firm grip
- Try to keep the arms within the boundary formed by the legs.
- The optimum position and nature of the grip depends on the circumstances and individual preference, but it must be secure.
- A hook grip is less fatiguing than keeping the fingers straight. If it is necessary to vary the grip as the lift proceeds, do this as smoothly as possible.

Don't jerk
- Carry out the lifting movement smoothly, keeping control of the load.

Move the feet
- Don't twist the trunk when turning to the side.

Keep close to the load
- Keep the load close to the trunk for as long as possible.
- Keep the heaviest side of the load next to the trunk.
- If a close approach to the load is not possible, try sliding it towards you before attempting to lift it.

Put down, then adjust
- If precise positioning of the load is necessary, put it down first, then slide it into the desire position.

Reproduced from *Manual Handling – Guidance on Regulation*, Health and Safety Executive.

Appendix 19

Basic tool list for structure building on adventure playgrounds

This list describes a basic selection of tools necessary to complete, safely and effectively, the range of tasks necessary in the building of timber play structures. Use 'middle of the range' tools. These tools will do the job properly and safely and suffer the rigours of life on an adventure playground over a reasonable period without breaking the bank. While it may be true that tools used on adventure playgrounds are more likely to suffer loss or damage it is a false economy, and a danger to workers and children, to use cheap, short-life tools.

Most adventure playgrounds will need to build on the basic kit described, buying some tools in larger quantities. This is especially true where large numbers of workers are to be involved in a major structure-building project over a short period of time, or where children are to be encouraged to use tools for dens and other building projects. You should consider carefully which tools it is safe for which children to use, you should not give children tools which are blunt or otherwise defective. This will place them in danger and frustrate them, as well as convincing them that you consider their projects to be a low priority.

Always use the correct tools for the job. Always look after your tools.

LIGHT TOOLS

Hammers	Claw hammer 16 oz (2)
	Club hammer $2\frac{1}{2}$ lb
Saws	Hand or crosscut saw 24 in, 6–8 ppi (2)
(ppi = points per inch)	Bow saw 30 in
	General purpose saw with adjustable blade
	Hacksaw 22 ppi, with spare blades
	Tenon saw

Drills	Universal ratchet brace 10 in sweep
	Centre bit $\frac{5}{8}$ in or 16mm (2)
	Scotch auger $\frac{5}{8}$ in or 16 mm(2)
	Hand drill
Spanners	Adjustable spanners e.g. 8 in and 10 in
	Ring spanner 20 mm to 22 mm
	Ring spanner 24 mm to 26 mm (2)
Chisels	1 in firmer chisel
	$1\frac{1}{2}$ in firmer chisel
Mallets	Joiners mallet (2)
Screwdrivers	Cabinet screwdriver 4 in
	Cabinet screwdriver 6 in
	Posi-drive screwdrivers

MEASURING TOOLS

Combination square (2)
Spirit level 30 in
Tape measure 3 m
Tape measure 30 m

MISCELLANEOUS

'Workmate' bench
'3-in-1' oil (tin)
Combination oil stone
Honing guide
Marking gauge ('mortice gauge')
WD40 'oil' (tin)
Knife (for cutting rope)
'Stanley' knife and spare blades
Working rope 12 mm rot-proofed sisal or polypropylene
Pulley block, single (2)
Pliers (combination)

HEAVY TOOLS

Pick axe with spike and blade (2)
Grubbing mattock
Spade (treading garden) $11\frac{1}{2}$ x $7\frac{1}{2}$
7 lb sledge hammer
Wrecking bars 24 in and 36 in

Nail bar
Shovel (2)
Aluminium extension ladder (safety harnesses and
ladder ties will be needed)
Wheel barrow (2)
Cold chisel 8 in

©PLAYLINK

POWER TOOLS (SOME EXAMPLES COMMONLY USED)

110 volt equipment is recommended and preferable. This does require a
transformer and extension cables in addition to the tools themselves.

If 240 volt equipment is used, a circuit breaker should always be used at the
power source.

Drills	Chuck must be large enough to take all envisaged bit sizes Auger bits: e.g. 16 mm long Shorter bits also advised
Saws*	Circular saw Reciprocating saw (or 'sabre'/'shark' saw) – multi-use Jig saw 'Alligator' saw Chain saw (electric) Chain saw (petrol) N.B. Chain saws require special training and safety clothing
Planer	Useful for smoothing rough edges
Sander	Useful for smoothing rough edges

110 volt transformer

110 volt extension cables (sufficient to run from transformer placed in building)

240 volt circuit breaker

*These are different types of saw, seek advice on which ones would be most suitable.

Most of the above are available for hire.

©PLAYLINK

Index

Page numbers in **bold** refer to figures.